Developmental and Stem Cell Biology in Health and Disease

Ahmed El-Hashash

Stem Cells, Regenerative Medicine and Developmental Biology Program, Children's Hospital Los Angeles, Keck School of Medicine and Ostrow School of Dentistry, University of Southern California, USA

advertisements or ideas contained in the Work.

Limitation of Liability:

In no event will Bentham Science Publishers, its staff, editors and/or authors, be liable for any damages, including, without limitation, special, incidental and/or consequential damages and/or damages for lost data and/or profits arising out of (whether directly or indirectly) the use or inability to use the Work. The entire liability of Bentham Science Publishers shall be limited to the amount actually paid by you for the Work.

General:

1. Any dispute or claim arising out of or in connection with this License Agreement or the Work (including non-contractual disputes or claims) will be governed by and construed in accordance with the laws of the U.A.E. as applied in the Emirate of Dubai. Each party agrees that the courts of the Emirate of Dubai shall have exclusive jurisdiction to settle any dispute or claim arising out of or in connection with this License Agreement or the Work (including non-contractual disputes or claims).
2. Your rights under this License Agreement will automatically terminate without notice and without the need for a court order if at any point you breach any terms of this License Agreement. In no event will any delay or failure by Bentham Science Publishers in enforcing your compliance with this License Agreement constitute a waiver of any of its rights.
3. You acknowledge that you have read this License Agreement, and agree to be bound by its terms and conditions. To the extent that any other terms and conditions presented on any website of Bentham Science Publishers conflict with, or are inconsistent with, the terms and conditions set out in this License Agreement, you acknowledge that the terms and conditions set out in this License Agreement shall prevail.

Bentham Science Publishers Ltd.
Executive Suite Y - 2
PO Box 7917, Saif Zone
Sharjah, U.A.E.
Email: subscriptions@benthamscience.org

**BENTHAM
SCIENCE**

CONTENTS

PREFACE

Stem cell research can be traced back for more than 20 years when scientists first isolated embryonic stem cells from mouse blastocysts, and a research article announcing the discovery of human embryonic stem cells emerged in 1998. Stem cell research field has rapidly expanded as new research and experience broaden our knowledge about different aspects of stem cell biology and applications. In the past decade, the stem cell field has grown very rapidly, and continues to be one of the most exciting aspects of biomedical research. This book brings together a number of topics that are related to stem cell development and behavior as well as stem cell applications in the treatment of different human diseases.

Although we could not hope to be comprehensive in the coverage of stem cells of different tissues, our main aim in compiling this book was to bring together a selection of the current progress in understanding stem cell biology and development as well as potential applications of stem cells in the treatments of different human diseases.

In preparing this book, we aimed at making it accessible to not only those working in stem cell biology field, but also to non-experts with a broad interest in stem cells and in human health. Our hope is that this book will be of value to all concerned with stem cell biology, development and application in medicine.

Ahmed El-Hashash
Children's Hospital Los Angeles
Keck School of Medicine and Ostrow School of Dentistry
University of Southern California
USA

List of Contributors

Ahmed A. Abd-Rabou
Hormones Department, Medical Research Division, National Research Center, Cairo, Egypt

Ahmed El-Hashash
Stem Cells, Regenerative Medicine and Developmental Biology Program, Children's Hospital Los Angeles, Keck School of Medicine and Ostrow School of Dentistry, University of Southern California, 4661 Sunset Boulevard, Los Angeles, USA

Ahmed R. N. Ibrahim
Stem Cells, Regenerative Medicine and Developmental Biology Program, Children's Hospital Los Angeles, Keck School of Medicine and Ostrow School of Dentistry, University of Southern California, 4661 Sunset Boulevard, Los Angeles, USA

Alhassen Wadah
Stem Cells, Regenerative Medicine and Developmental Biology Program, Children's Hospital Los Angeles, Keck School of Medicine and Ostrow School of Dentistry, University of Southern California, 4661 Sunset Boulevard, Los Angeles, USA

Azza El Amir
Zoology Department, Faculty of Science, Cairo University, Giza, Egypt

Deshna Majmudar
Stem Cells, Regenerative Medicine and Developmental Biology Program, Children's Hospital Los Angeles, Keck School of Medicine and Ostrow School of Dentistry, University of Southern California, 4661 Sunset Boulevard, Los Angeles, USA

Elham M. Youssef Elabd
National Research Center, Suez University, Suez, Egypt

Hadeer A. Aglan
Hormones Department, Medical Research Division, National Research Center, Cairo, Egypt

Haifen Huang
Stem Cells, Regenerative Medicine and Developmental Biology Program, Children's Hospital Los Angeles, Keck School of Medicine and Ostrow School of Dentistry, University of Southern California, 4661 Sunset Boulevard, Los Angeles, USA

Hani S. Hafez
Suez University, Suez, Egypt

Hanaa H. Ahmed
Hormones Department, Medical Research Division, National Research Center, Cairo, Egypt

Jesse Garcia Castillo
Stem Cells, Regenerative Medicine and Developmental Biology Program, Children's Hospital Los Angeles, Keck School of Medicine and Ostrow School of Dentistry, University of Southern California, 4661 Sunset Boulevard, Los Angeles, USA

John Ku	Anatomy Department, Faculty of Medicine, Mansoura University, Mansoura, Egypt Rehabilitation Science Department, College of Applied Medical Sciences, King Saud University, Riyadh, KSA
Manal E. Elsawaf	Faculty of Medicine, Tanta University, Tanta , Egypt
Marwa E. Elgayyar	Stem Cells, Regenerative Medicine and Developmental Biology Program, Children's Hospital Los Angeles, Keck School of Medicine and Ostrow School of Dentistry, University of Southern California, 4661 Sunset Boulevard, Los Angeles, USA
Mohamed Berika	Anatomy Department, Faculty of Medicine, Mansoura University, Mansoura, Egypt Rehabilitation Science Department, College of Applied Medical Sciences, King Saud University, Riyadh, KSA
Marwa Adel HasbySaad	Faculty of Medicine, Tanta University, Tanta , Egypt
Nahla M. Shoukry	Suez University, Suez, Egypt
Noha M. Osman	National Research Center, Suez University, Suez, Egypt Department of Molecular and Computational biology, University of Southern California, Los Angeles, USA Washington State University, University of Southern California, Pullman, USA
Safia Gilani	Stem Cells, Regenerative Medicine and Developmental Biology Program, Children's Hospital Los Angeles, Keck School of Medicine and Ostrow School of Dentistry, University of Southern California, 4661 Sunset Boulevard, Los Angeles, USA
Salaheldin S. Soliman	Stem Cells, Regenerative Medicine and Developmental Biology Program, Children's Hospital Los Angeles, Keck School of Medicine and Ostrow School of Dentistry, University of Southern California, 4661 Sunset Boulevard, Los Angeles, USA
Sameh Elshahawy	Stem Cells, Regenerative Medicine and Developmental Biology Program, Children's Hospital Los Angeles, Keck School of Medicine and Ostrow School of Dentistry, University of Southern California, 4661 Sunset Boulevard, Los Angeles, USA
Sara M. Abdo	Chemistry Department, Faculty of Science, Helwan University, Cairo, Egypt
Wajeet Nabil	Zoology Department, Faculty of Science, Cairo University, Giza, Egypt

Developmental and Stem Cell Biology in Health and Disease

Editor: Ahmed El-Hashash

ISBN (eBook): 978-1-68108-219-6

ISBN (Print): 978-1-68108-220-2

Developmental and Stem Cell Biology in Health and Disease

2

CHAPTER 1

Stem Cells, Developmental Biology and Reparative/ Regenerative Medicine: Tools and Hope for Better Human Life

Ahmed El-Hashash[*]

Stem Cells, Regenerative Medicine and Developmental Biology Program, Children's Hospital Los Angeles, Keck School of Medicine and Ostrow School of Dentistry, University of Southern, California, USA

Keywords: Stem Cells, Developmental Biology, Regenerative Medicine.

The development of all humans begins after the union of male and female gametes or germ cells during fertilization or conception. The fertilized egg or zygote is a large diploid cell that is the beginning, or primordium, of a human being. This fertilized egg undergoes rounds after rounds of both highly organized and tightly controlled cell division until it comprises many billions of stem and lineage specific cells that have self-renewal and self-repairing capabilities and form the human body. These processes are studied in a branch of science called developmental biology that explores how organisms develop and progress. As a developmental and stem cell biologist, who have investigated the mechanisms of organogenesis in different tissues and organs such as neural crest cells, placenta, kidney and lung, it becomes clear to me that, if we can understand these normal and fundamental mechanisms of developmental biology, then correcting abnormalities caused by congenital defects, repairing the injured tissues and even generation of functional whole organs from stem cells should be theoretically

[*] **Correspondence author Ahmed El-Hashash:** Developmental Biology, Regenerative Medicine and Stem Cell Program, Saban Research Institute, Children's Hospital Los Angeles, 4661 Sunset Boulevard MS 35, Los Angeles, California 90027, USA; Tel: 323-361-2764, 323-361-2258; Fax: 323-361-3613; E-mail: aelhashash@chla.usc.edu

achievable. The stem cell field has grown very rapidly over the past decade, and continues to be one of the most exciting aspects of biomedical research. Stem cell is a fast growing filed of research, with an astonishing annual growth rate of 77% since 2008. The volume of research output, and thus publication, has therefore increased significantly in all areas of stem cell research. By now, the first functioning whole organ, thymus, has been generated in the laboratory, and the first in vitro fertilized human baby girl has children of her own. Research is currently underway in different laboratories worldwide to generate other functioning whole organs such as the intestine and kidney.

Embryonic stem cells (ESCs) were isolated from mouse blastocysts by scientists in 1981, while human ESCs were first reported in 1998. Currently, adult-derived stem cells (ASCs) are also a favorite subject of intensive research investigation. Recently, ESCs are almost routine in the face of more advances in the medical field. More recent advances show the possibility to drive fully differentiated cells back-wards towards a more embryonic like state of induced pluripotency. This occurs by means of as few as four factors, and represents a major scientific discovery. Moreover, it has been recently shown that several classes of stem-like cells, which are originating from different mesenchymal compartments of the body such as the amnion, marrow, amniotic fluid, adipose, exert promising exert therapeutic effects in some inflammatory and fibrotic and diseases. In addition, neural stem cells can be programmed to selectively travel and attack inaccessible brain tumors. Furthermore, recent identification of endophenotypes, or latent risk factors, for certain types of aggressive cancers my eventually lead to designing novel strategies for cancer treatments. Together, these recent discoveries could identify the next generation of treatments emerging from our scientific discoveries.

Scientists worldwide are applying new stem cell discoveries to the betterment of human diseases, which has brought forth much hope for better human life. The branch of translational research in tissue engineering and molecular biology, which takes advantage of rapid progress in our understanding of stem cell biology during development and adulthood, is called regenerative medicine. The hope for cures of different diseases has prompted different countries worldwide to invest in stem cell research and regenerative medicine.

The United States, for example, plays a critical role in stem cell research, likes several other countries in the world. Many countries in Europe and Asia, in addition to Canada and Brazil, have leading centers for stem cell research and regenerative medicine. These research centers have significantly expanded the scope of stem cell research and their applications in the treatment of different human diseases.

This book contains a global collection of monograph essays from collaborating research scientists at different research institutes and countries. They describe exciting progress in basic stem cell biology and regenerative medicine, including the application of stem cell therapy in different human diseases.

CHAPTER 2

The First Morphogenetic Events During Mammalian Development

Manal E. Elsawaf*

Faculty of Medicine, Tanta University, Egypt

Abstract: The question of how could a single cell develop into an animal has been asked for centuries. Since the seventeenth century, epigenesis and preformation theories have been two persistent ways seeking to explain the development of individual organic form. Nowadays, it is proved that both zygote's genome and cytoplasmic determinants control development. Cell division, cell differentiation and morphogenesis then take place. To build an animal's body, fertilization is the first step where a diploid number of chromosomes is restored. Fusion of egg and sperm activates the egg. Three stages then follow; cleavage, gastrulation and organogenesis. Cleavage pertains to the repeated mitotic division of a zygote into smaller cells, blastomeres. More cleavage results in a solid ball of cells called morula. With further cleavage a hollow ball of cells, the blastocyst, is produced. Gastrulation is a necessary event in developing a multicellular animal. During this process, the embryonic cells are rearranged to form a three layered embryo. Accordingly, cells acquire new positions enabling them to interact with cells that were initially far away from them. Many inductive interactions then occur to start neurulation and organogenesis. Early in vertebrate organogenesis, the notochord which forms in mesoderm leads to neural plate induction from the covering ectoderm. This neural plate forms the neural tube that will become the central nervous system. All other organs develop from folds, splits and condensations of cells. Thorough understanding of early mammalian development has initiated the era of embryonic stem cell generation and its use in medicine.

Keywords: Acrosome reaction, Amniotic cavity, Blastocyst, Capacitation, Cleavage, Compaction, Ectoderm, Endoderm, Epigenesis, Fertilization, Folding, Gastrulation, Implantation, Mammalian development, Mesoderm, Morphogenesis,

---

* **Correspondence author Manal E. Elsawaf:** Faculty of Medicine, Tanta University, Egypt; Tel: 00201287319509-0020403280868; E-mail: elsawafmanal@yahoo.com

Neurulation, Notochord, Organogenesis, Pluripotency, Preformation, Primitive streak, Stem cells, Yolk sac.

INTRODUCTION

The question of how could a single cell develop into an animal has been asked for centuries. In the late 17th century, the scientists suggested the preformation theory. They suggested that the zygote contains an invisible miniature infant which becomes unfolded and expanded, and hence becomes larger during development. On the contrary, Aristotle, in his book "on the generation of animals" originated the theory of epigenesis. He proposed gradual emergence of form from un-form. Nowadays, it is proved that both zygote's genome and cytoplasmic determinants control development [1, 2].

Embryology is a part of a broader science, developmental biology, which deals with the study of emergence of a new individual from fertilized egg. Most of our knowledge in this field is collected from studying mice embryos and to a less extent other animal species including human. During the past decades, researches in the field of developmental biology have grown up dramatically through the extensive knowledge of whole genome sequences from different organisms, the advancement of imaging techniques, and thorough understanding of the role of stem cells in development and regeneration of organs and tissues. These studies have increased our knowledge regarding the cellular origins of organs and tissues and provided more understanding to the mechanisms coordinating cellular reorganization in morphogenesis. Morphogenesis in a complex multicellular organism means its ability to recruit, reorganize, and reshape groups of cells to form functionally specialized tissues and organs. The right cells must be relocated in the accurate place and interact at the right time in order to produce a tissue or an organ specified for a particular function. Monitoring cell fates provides valuable information about the origin of different organs and tissues. The behavior of cells appears to be a destiny rather than being a chance or a choice. When cells do not obey the rules, birth defects may result in. Moreover, prenatal experiences in combination with molecular and cellular factors may determine the potential to develop certain adult diseases.

Mammalian development from fertilization till the first steps of organogenesis is the scope of this chapter. Although the duration of intrauterine development of mammalian embryos varies in length from 16 days in golden hamster up to 15-17 months in the rhinoceros or the cachalot, morphogenetic changes in the early phases of embryonic development are similar in different mammals [3]. These changes occur in an uninterrupted, overlapping developmental stages based on external and morphological criteria irrespective to age or length of the embryo [4]. Moreover, several reports suggested that the mechanisms of lineage specification may differ in a great extent among different species of mammals [5 - 8]. Thus, the aim of this chapter is to describe the early morphogenetic changes that happen in all mammalian embryos. We will focus on the specific characters confined to mammals which characterize the development of their embryos from other species. These characters can be summarized as follows:

1. Fertilization is internal.
2. The zygote produces two different groups of cells with different fates; embryonic and extraembryonic tissues.
3. The zygote is implanted into the maternal uterus.
4. The embryo gets his nourishment through his mother.

Based on these characters, early mammalian development, from fertilization till the start of organogenesis, can be classified into three periods:

1. Mammalian pre-implantation development.
2. Mammalian development during implantation.
3. Early mammalian post-implantation development.

Study of early development of mammalian embryos and thorough understanding of cell behavior open the door towards the era of embryonic stem cell generation and its clinical application. This current chapter is going to throw light on this issue.

1. MAMMALIAN PRE-IMPLANTATION DEVELOPMENT

The mammalian embryo is characterized by being attached to the maternal uterine epithelium during most part of its development. Accordingly, the mammalian zygote produces two different groups of cells: embryonic and extraembryonic tissues. The extraembryonic tissues are formed during the period of pre-implantation development and are required for implantation of the embryo into the uterus. This period starts with fertilization and formation of the zygote that is followed by multiple mitotic divisions called cleavage till the formation of blastocyst [9].

Fertilization

Building up an animal's body is switched on by fertilization in which union of two radically different - looking haploid cells, sperm and egg, occurs. When the ovulated egg and ejaculated sperm meet each other in the uterine tube, subsequent multi-step events occur to form a zygote. However, prior to formation of a zygote, many biochemical, physiological and morphological changes must happen in both male and female gametes.

Changes in Male Gamete

At coitus, millions of sperms are deposited in the female vagina. However, only a small percentage of these sperms ascend to the uterus. Muscular contractions of the uterus and uterine tube draw the sperm from the cervix. Little movement is shown by sperm's own propulsion. Researches have been done to study the essential genes that govern the journey of the sperm towards the uterine tube. Many observations have postulated that there is a recognition system between the sperm and the uterotubal junction which allows sperm entry to the uterine tube [10 - 12]. In addition, some evidence indicates that the arrival of sperms towards the egg side does not occur by a chance but rather by a chemo- attractant released from the follicle cells surrounding the egg [13]. In most of the studied mammalian species, including man, ejaculated spermatozoa are not able to bind to an egg and fertilize it immediately. They require a period of residence in the female genital tract through which they undergo multiple biochemical and physiological changes to become functionally competent cells. This period of conditioning is referred to as ***capacitation*** [14, 15]. Although numerous studies discussed this process, the actual mechanisms that govern capacitation have been obscure [16, 17]. During this process, the plasma membrane covering the sperm's acrosome loses a glycoprotein coat as well as seminal plasma proteins. Only capacitated spermatozoa can pass freely through corona cells and interact with the extracellular glycocalyx egg coat, the ZP (Zona Pellucida) [13 - 15].

The binding capacity of sperm to the ZP is species- specific [18]. After capacitation, the sperm must undergo a process called ***acrosome reaction*** prior to fertilization. The acrosome is a subcellular organelle located at the top of the sperm head. It contains multiple lytic enzymes and zona pellucida binding proteins. Multiple mechanisms take place to allow fusion of both the plasma membrane and outer acrosomal membrane, and hence the release of acrosomal contents. The release of these enzymes helps sperm to pass through the zona and come in contact with the plasma membrane of the oocyte [19, 20]. Finally, sperm binds to the plasma membrane and fuses with it.

Changes in Female Gamete

Sperm entry initiates Ca^{2+} oscillations in the oocyte. This sharp rise in Ca^{2+} activates the egg and stimulates the second meiotic division to be completed [21]. Activation of the egg also initiates the ***cortical reaction*** in which cortical granules release their contents outside the egg. As a result, the egg plasma membrane does not fuse with any other sperm that have penetrated the ZP. Further, the permeability of the ZP is altered and the species-specific receptors for sperms on the zona surface are inactivated. These changes act as a block to ***polyspermy*** [22] (Fig. **1**).

Upon fertilization, the definitive oocyte is formed while the second polar body is extruded. The nucleus of the definitive oocyte is known as the female pronucleus while that of the spermatozoon swells to form the male pronucleus. Both male and female pronuclei approach each other and then lose their nuclear envelopes. During this period, both nuclei must replicate their DNA. After DNA replication, chromosomes arrange on the spindle. A single diploid cell, the zygote, has been formed now and is ready for its first mitotic division [13].

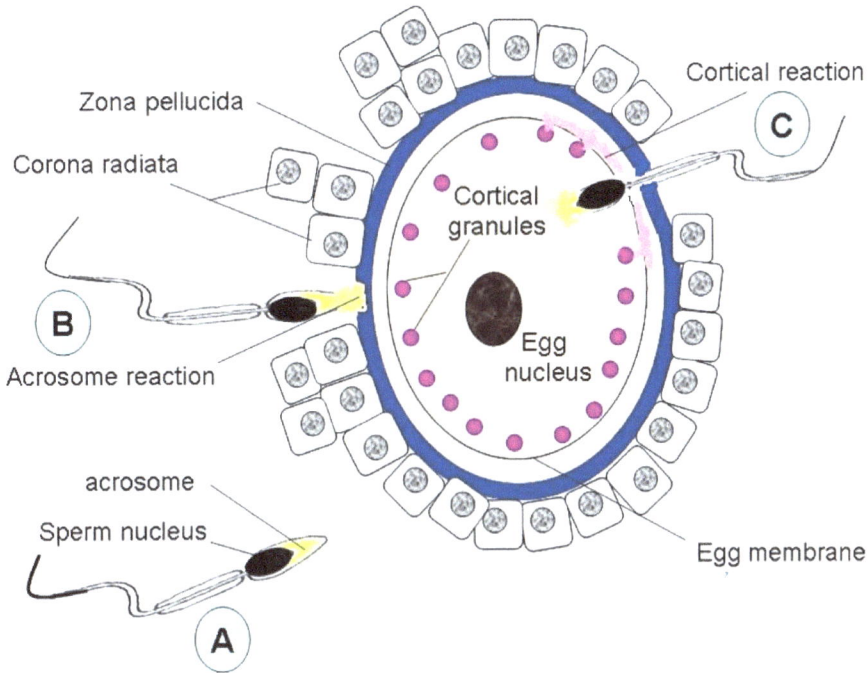

Fig. (1). Fertilisation: **A**. Spermatozoa meat the egg. **B**. Spermatozoa pass through corona radiata cells, and fuse with zona pellucida leading to acrosome reaction. **C**. One spermatozoon penetrates the egg membrane leading to cortical reaction.

Cleavage

Cleavage pertains to the repeated mitotic division of a zygote into smaller cells called blastomeres. Although, the number of cells within the embryo increases, its total cellular volume remains the same. The cytoplasm is distributed approximately equally among daughter blastomeres. More cleavage results in the formation of a solid ball of cells called a morula. Further cleavage produces a hollow ball of cells called the blastocyst. This process occurs post fertilization, while the conceptus migrates down the uterine tube. Because the maternal stores are of small amount, activation of the zygotic genome of the embryo must occur early at the end of the 1-cell stage by a minor burst. This is followed by a major burst at the end of the 2-cell stage with degradation of the maternal mRNAs [23 - 25].

Up to the formation of eight cells, blastomeres are spherical and touch each other loosely with a symmetrically organized cytoplasm and a centrally located nucleus. Then, a process known as ***compaction*** proceeds. During this process, blastomeres flatten on each other. Thus, intercellular contact between blastomeres increase and adherens junctions appear between cells [26, 27]. The cytoplasm of each blastomere is reorganized into a highly asymmetric phenotype, while their nuclei migrate towards the centre of the embryo, and their surface microvilli are redistributed to the exposed surface. Thus, all the eight cells forming the embryo are polar with their apices facing outwards and their basolateral surfaces facing internally. However, with further division and formation of the 16-cell embryo, two cell populations are identified; polar and apolar. The polar cells are the outer or superficial cells which differentiate an epithelial phenotype, the ***trophectoderm***. The apolar cells are the inner cells which form the ICM (***inner cell mass***). This means that compaction is crucial for generating cell diversity so early in the embryo [28, 29]. Thus, segregation of cell diversity takes place in the 16-cell embryo before blastocyst formation. After this initial segregation toward trophectoderm and ICM, their identities are governed by a small number of transcription factors [30, 31].

In the next division (16 to 32 cells), cells divide leading to formation of both polar and apolar populations which join the trophectoderm and ICM respectively. After formation of 32- cell embryo, the outer polar cells (trophectoderm) permit transport of fluid in an apical to basal direction to create a cavity called ***blastocoel***. With further division to 64 cells, this cavity enlarges to convert the morula to blastocyst. The inner cell mass cells appear at one pole and form the embryoblast, while the trophoblast (outer cells) forms the wall of the blastocyst. In late blastocyst, the trophoblast is identified as polar trophoblast that covers the embryoblast and mural trophoblast that surrounds the blastocyst cavity [32 - 34] (Fig. **2**).

Thus during early stages, cells become fully devoted to either trophectoderm or inner cell mass. Accordingly, the trophectoderm is considered the first differentiated tissue in the embryo. This tissue is necessary for the process of implantation of the embryo and later provides a contribution to the placenta. During early stages also, cells of the ICM begin to be arranged into two layers of cells namely ***primitive endoderm*** or ***hypoblast*** and ***epiblast*** cells. Cells that face the blastocoel form the primitive endoderm which later becomes the tissues of the yolk sac. The other deep cells form the epiblast which is the progenitor for all cells in the embryo [35].

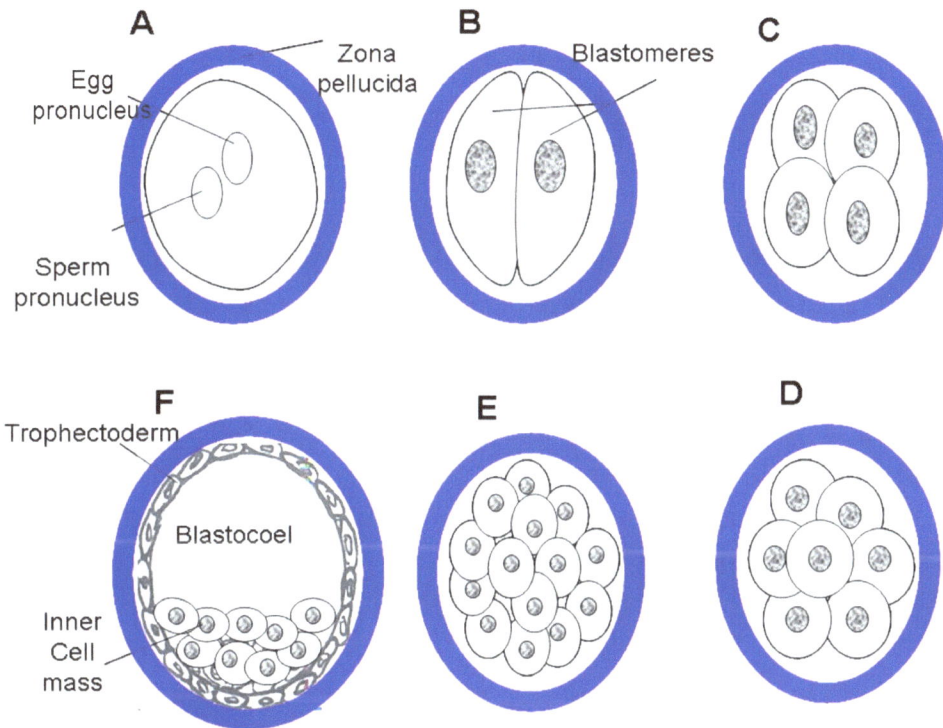

Fig. (2). Cleavage: A. Fertilized ovum. **B**. Two-cell stage. **C**. Four-cell stage. **D**. Eight-cell stage. **E**. Morula. **F**. Blastocyst.

At the blastocyst stage, a polarity axis can be recognized where the ICM forms the embryonic pole while the trophectoderm facing the blastocyst cavity forms the abembryonic pole. Later, this embryonic – abembryonic axis represents the dorsoventral axis of the post- implantation embryo. Thus, at or even before implantation, the developing embryo shows both established dorsoventral axis and a bilaminar arrangement of its ICM cells [33, 36].

2. MAMMALIAN DEVELOPMENT DURING IMPLANTATION

Implantation is a necessary event during mammalian development which allows mammals to provide nourishment and protection for their embryos. The eventual objective of implantation in all mammals is to enable the conceptus to attain the maternal blood supply. The onset of implantation differs in different animals. It occurs 4 days post coitum in mice and 9 days in humans while it needs about 30 days after fertilization in cows [37]. The physiological mechanisms that govern the process of implantation and the morphological changes occurring in the embryo during this process can be easily studied and observed in different animal models [38 - 40].

Although the morphological changes of the embryo during implantation are similar in all mammals, the blastocyst–uterine cell interactions differ in different species. Implantation in mammals has been classified into centric, eccentric and interstitial. The first category, centric implantation, occurs in cases where the blastocyst grows and comes into contact with uterine epithelium but without penetrating through it. This type is demonstrated in rabbits, dogs, cows, pigs, and sheep. In eccentric implantation, as in mice, rats and hamsters, the uterine epithelium shows an invagination which surrounds the trophoblast. In interstitial type of implantation, the trophoblast invades through the uterine epithelium and becomes imbedded into the endometrium. This type appears in humans and guinea pigs [41].

Implantation involves multiple events starting from attachment of the blastocyst to the endometrial surface till the establishment of the utero- placental circulation. As a result of this blastocyst-endometrium interaction, morphological changes occur in both. At the site of implantation, the endometrium undergoes decidual reaction in which the endometrial stromal cells proliferate and differentiate into morphologically different decidual cells. The endometrium becomes extremely vascular and edematous. The endometrial glands become elongated and tortuous and secrete abundant glycogen and mucus. Soon after, the decidual reaction occurs throughout the endometrium [42]. There is evidence that decidual cells secrete insulin-like growth factor binding protein1 and prolactin which may be taken up by the trophoblast. These secretions probably play a role in the maintenance and growth of the conceptus in the early part of postimplantation development [33]. Meanwhile, multiple changes occur in the blastocyst involving the trophoblast and the ICM as well as the blastocyst cavity.

For implantation to begin, the blastocyst hatches from its ZP, with the help of an enzyme similar to trypsin [33]. Hatched blastocyst expands and cells of the ICM are arranged into two layers to form a flat bilaminar disc. One layer is formed of small cuboidal cells roofing the blastocyst cavity (primitive endoderm or hypoblast), while the other layer consists of high columnar cells away from the blastocyst cavity (the epiblast). Soon, the epiblast layer acquires a small cavity known as AC (***amniotic cavity***). The epiblast cells which intervene between the AC and the trophoblast is termed ***amnioblast***. Thus, the bilaminar disc lies between the amniotic cavity dorsally and the blastocyst cavity ventrally. Blastocyst approaches to the uterine wall to start the process of implantation. The trophoblast cells over the embryoblast pole are oriented towards the endometrium and adhere to it. L selectin on the trophoblast cells mediates the initial attachment of the blastocyst to the carbohydrate receptors on the endometrial epithelium. Moreover, the trophoblast cells express integrins for further attachment and invasion of the blastocyst to the uterine wall. Interaction of integrins with laminin molecules of the extracellular matrix enhances attachment, while integrins interaction with fibronectin promotes migration. This means, implantation is a mutual trophoblastic and endometrial action [13].

Contact with the uterine endometrium induces the trophoblast at the embryonic pole to differentiate into an inner layer of mononucleated cells called ***cytotrophoblast*** and an outer multinucleated layer called ***syncytiotrophoblast.*** Cells of the cytotrophoblast layer continue to divide and migrate to the syncytiotrophoblast, where they lose their cell membranes and fuse together [13, 33]. Vacuoles appear in the syncytium at the embryonic pole and fuse together to form large lacunae. Meanwhile, cells of the syncytiotrophoblast invade the endometrial stroma and erode the lining of the maternal blood sinusoids. Accordingly, the maternal blood sinusoids become continuous with the syncytial lacunae leading to flow of maternal blood into the lacunar system and establishment of the uteroplacental circulation [43]. As implantation progresses, the expanding syncytiotrophoblast gradually envelops the blastocyst.

Meanwhile, the hypoblast gives rise to a group of flattened cells which form an *exocoelomic* or *Heuser's membrane*. This membrane lines the cavity of the *primitive yolk sac* while the hypoblast roofs it. Another type of cells appears between the cytotrophoblast and the lining of the primitive yolk sac, the*extraembryonic mesoderm*. This layer of cells is believed to originate in humans from the hypoblast/ primary yolk sac. In the mouse embryo, it arises from the caudal end of the incipient primitive streak and in addition, the trophoblast may contribute cells as well [34]. Then, the extraembryonic mesoderm spreads between the trophoblast externally and the amnion and the exocoelemic membrane internally. This is followed by the appearance of cavities in the extraembryonic mesoderm which fuse together to form the extraembryonic cavity *(chorionic cavity)*. The chorionic cavity surrounds both the primitive yolk sac and the amniotic cavity except at the connection between the embryonic disc and the trophoblast, *the connecting stalk* (future umbilical cord). Cytotrophoblast cells proliferate and form cellular columns which penetrate through the syncytiotrophoblast to form *primary villi*. Further, the hypoblast gives rise to cells which migrate along the inside of the exocoeleomic membrane with formation of a new cavity within the chorionic cavity known as the *definitive* or *secondary yolk sac* [13, 33] (Fig. **3**).

Thus, when the implantation process is completed, the implanting embryo contains three cavities where the blastocyst cavity is now termed the chorionic cavity. This cavity is large and surrounds two much smaller cavities; the amniotic cavity and the secondary yolk sac. Later, a fourth cavity protrudes into the connecting stalk as a caudal hypoblastic diverticulum and is termed the *allantois*.

3. EARLY MAMMALIAN POST-IMPLANTATION DEVELOPMENT

When the process of implantation is completed, the embryo undergoes the process of gastrulation with formation of the three germ layers. Two crucial events then occur to induce organogenesis; the formation of the notochord and folding of the embryo. Meanwhile, further development of the trophoblast occurs where the central part of the primary villi is invaded with mesoderm to form the secondary villi. Later, some mesodermal cells, in the center of the secondary villi, differentiate into blood vessels to form the tertiary villi.

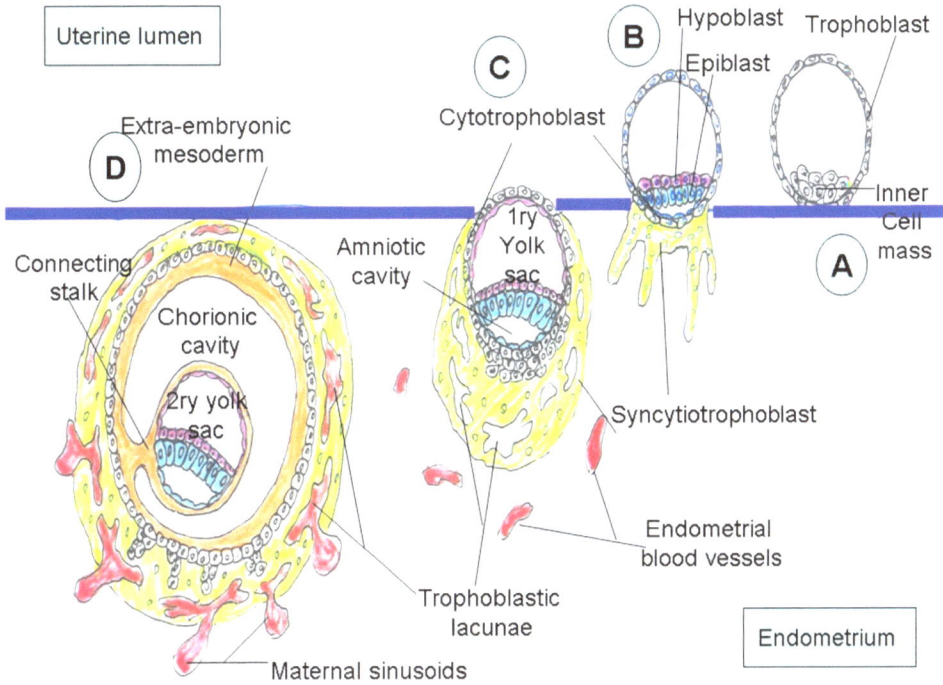

Fig. (3). Implantation (interstitial type): **A**. Blastocyst adheres to the endometrium. **B**. Trophoblast differentiates into cytotrophoblast and syncytiotrophoblast and begins to penetrate through endometrium. Inner cell mass segregates into hypoblast and epiblast. **C**. The bilaminar embryonic disc separates between the amniotic cavity and primitive yolk sac. Multiple trophoblastic lacunae appear in the syncytiotrophoblast. The endometrial blood vessels are dilated. **D**. The trophoblastic lacunae join the maternal sinusoids with establishment of utero-placental circulation. Three cavities are present: amniotic cavity, secondary yolk sac and chorionic cavity. Note the extra-embryonic mesoderm and primary villi.

Gastrulation

After implantation of blastocyst, morphogenetic movements begin to occur to arrange the embryo into three germ layers; ectoderm, mesoderm and endoderm. Cells in these germ layers are the primitive building blocks for formation of organ rudiments. Thus, the cells in the different layers will ultimately have different developmental fates. These morphogenetic movements involve shape changes, migration, and intercalation of individual cells and groups of cells. As a result of these movements, cell populations from different germ layers that are required for formation of different organs or body parts are brought together. This approximation of tissues facilitates inductive interactions that are essential for lineage specification and tissue patterning [34, 44, 45].

The first step in the process of gastrulation is the genesis of the ***primitive streak*** on the dorsal epiblastic aspect. The primitive streak is considered the site of the organizer cells. It appears at the caudal region of the embryonic disc as a collection of pluripotent cells oriented along its long axis in the median position. Its appearance determines the future craniocaudal axis of the embryo. The formation of primitive streak is induced by its underlying visceral hypoblast. The cephalic end of the streak is called ***primitive node*** which surrounds the ***primitive pit***. At the primitive streak, the epiblast cells show a period of active proliferation [13, 46].

Some epiblast cells migrate towards the primitive streak where they become flask shaped and separate from the epiblast to lie underneath it in a process called ***invagination*** or ***ingression***. Migration of epiblast cells is controlled by FGF8 (fibroblast growth factor 8). FGF8 downregulates the protein, E-cadherin, which normally binds the epiblast cells. In addition, it regulates Brachyury T expression to control cell specification into the mesoderm. These ingressed cells either replace the hypoblast to form the embryonic endoderm or locate between the epiblast and the newly formed endoderm to form the intraembryonic mesoderm. The other cells in the epiblast layer form the ectoderm. Thus, great evidences denote that the epiblast cells produce all the germ layers that will differentiate to all tissues and organs in the developing embryo [13, 47] (Fig. **4A**).

Studies of cell destiny have suggested that the fate of epiblast cells is determined at or even before the time of their ingression into the primitive streak, denoting the high significance of their ingression for future differentiation. Many cells that ingress through the primitive streak and lie between epiblast and hypoblast start to move laterally and cranially. Some of these cells reach to the disc margin and even contact with the extraembryonic mesoderm. Passage through the streak is specified according to position. Cells that pass through the primitive node differentiate into the prechordal mesenchyme, the notochord, the definitive endoderm, and medial halves of the somites. The lateral halves of the somites arise from the most cranial aspect of the streak. The middle region of the streak gives the lateral plate mesoblast. The caudal aspect of the streak produces the primordial germ cells, whereas the terminal caudal portion contributes cells to the extraembryonic mesoblast.

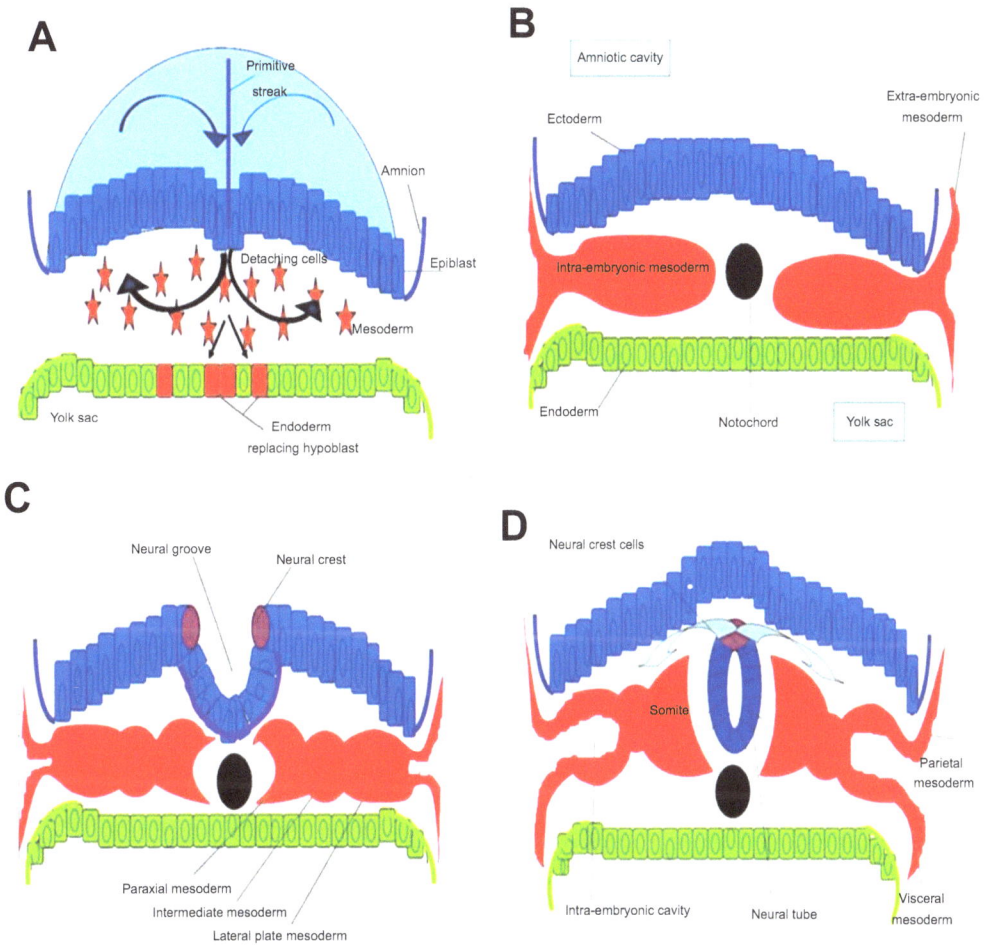

Fig. (4). Gastrulation and neurulation : **A** . Gastrulation: epiblast cells ingress through the primitive streak to replace the hypoblast cells with endoderm and to form the intraembryonic mesoderm. The remaining epiblast is considered ectoderm. **B** . The trilaminar embryonic disc is formed of the three primary germ layers; ectoderm, mesoderm and endoderm. Note the notochord in the mesodermal layer. **C** . Induction of neural groove and segmentation of mesoderm into paraxial, intermediate and lateral plate mesoderm. Note the neural crest. **D** . The neural tube is formed. The paraxial mesoderm differentiates into somites and the lateral plate mesoderm shows the intra-embryonic cavity inside.

The remaining cells in the epiblast layer produce the neural plate and surface ectoderm of the embryo [33]. The ectoderm and endoderm are tightly adherent at both cranial and caudal ends of the embryonic disc to form the oropharyngeal and cloacal membranes respectively. When the cloacal membrane is formed, the allantois appears as a diverticulum from the posterior wall of the yolk sac projecting into the connecting stalk [13, 34].

The first migrating cells through the node in the midline move cephalically to form the ***prechordal plate***. It is seen between the oropharyngeal membrane and rostral part of the notochord. It is described as a temporary collection of cells which underlies the neural plate and it is composed of cells that are similar to, or larger and more spherical than, the ingressing endodermal cells [48]. First, the prechordal plate is formed of many cell layers that extend along the long axis of the embryonic disc. Then, the lateral cells of the plate migrate laterally and the height of the plate reduces. Later, these migrating cells form bilateral premandibular mesenchymal condensations [33]. Prechordal plate may also contribute to the oropharyngeal membrane [34]. In addition, the prechordal plate plays an important role in inducing forebrain [13, 34].

Formation of the Notochord

Some epiblast cells invaginate through the primitive node and go towards the prechordal plate along the midline. These cells are called prenotochordal cells that will form the notochordal plate. Because both the notochordal cells and the endodermal cells ingress at the same time, they are initially mixed together. However, when the hypoblast is replaced by endoderm, the notochordal cells proliferate and leave the endoderm to form the definitive notochord. For a short time during development of the notochord, the amniotic cavity and the yolk sac are communicated through an opening in the primitive pit termed ***neuroenteric canal***. The primitive notochord helps in maintenance and development of the neural floor plate and has a role in induction of motor neurons. The definitive notochord is important for development of the axial skeleton [13, 34, 49].

Establishment of the Body Axes

When the primitive streak is formed along the midline of the caudal part of the dorsal aspect of the epiblast, the embryo becomes bilaterally symmetric. Three body axes; anteroposterior, dorsoventral, and left-right, can be identified so early. However, establishment of these axes are signalled by specific cells which express many transcription factors and secreted factors. In mouse embryos, cranial patterning actually occurs before the appearance of the primitive streak, as a result of signalling from the anterior visceral endoderm. This area produces genes crucial for head formation [1, 13, 34].

Changes of the Shape of the Embryonic Disc

Initially, the embryo appears as a flat rounded disc which gradually elongates and becomes pear shaped with its cephalic end broader than its caudal end. Growth of the cephalic end is because of the continuous migration of the cells through the primitive streak. Later, the primitive streak itself regresses, shrinks and disappears. The germ layers of the cranial part of the embryonic disc begin specific differentiation earlier than the caudal part. Thus, while the process of gastrulation is still proceeding in the caudal embryo, its cephalic part is differentiating, causing the embryo to develop cephalocaudally [13].

Organogenesis

Organogenesis or the embryonic period is the period when the three germ layers of the embryo differentiate into specific tissues and organs. It has been reported that both humans and mice encounter similar embryogenesis process and share most of their genes. However, the mechanisms that regulate these genes are so different that they lead to evident variations between species [50]. One of the major morphological changes in this stage is conversion of the flat embryonic disc into a somewhat cylindrical form by a process termed as ***folding***.

Embryonic Folding

Embryonic folding occurs early during the period of organogenesis as the embryo rapidly grows and folds around the yolk sac, and thus creates the basic vertebrate body form, called the ***tube- within- a tube body plan***. The main force responsible for embryonic folding is the differential growth of different portions of the embryo. As the cephalic and caudal ends curve around the ends of the yolk sac, the embryo becomes C-shaped with his head and tail almost touching. Meanwhile, the lateral margins of the embryonic disc fold around the sides of the yolk sac to form the ventral aspect of the embryo. The ventral body wall closes completely except for the umbilical region where the connecting stalk and yolk sac duct remain attached. This lateral folding encloses the primitive gut which later produces the digestive tract. Thus as a result of folding, each of the endodermal, mesodermal, and ectodermal layers of the embryonic disc fuse to the corresponding layer on the opposite side creating a tubular three dimensional body form [1, 34].

Morphogenetic Changes of the Ectodermal Germ Layer

Once formation of the definitive endoderm and intraembryonic mesoderm is complete, epiblast cells no longer ingress through the primitive streak. Thus, the remaining epiblast contributes the ectoderm, which quickly differentiates into the central neural plate and peripheral surface ectoderm. The ectoderm overlying the notochord and prechordal mesoderm is thickened to form the neural plate. Neural plate induction is regulated by inactivation of the growth factor BMP4 (bone morphogenetic protein 4). Thus, the initial step in the process of ***neurulation*** is the neural plate formation. Absence of inactivation of BMP4 directs the ectoderm towards the formation of epidermis and the mesoderm towards the formation of intermediate and lateral plate mesoderm. In general terms, the ectodermal germ layer differentiates into the epidermis, the nervous system, and the sensory epithelium of ear, nose and eye. In addition, it produces the enamel of the teeth, the hair, the nails and some glands as subcutaneous glands and mammary glands [13].

After formation of the neural plate, multiple morphogenetic events occur leading to formation of the ***neural tube***. The lateral edges of the neural plate become elevated to form the ***neural folds*** while its midregion becomes depressed to form the ***neural groove***. The two neural folds approach each other to fuse forming the neural tube (Fig. **4B, C, D**). The fusion level differs among species. In the human embryo fusion starts at the cervical region and proceeds cranially and caudally. Neural tube is connected by openings called ***neuropores*** at its cranial and caudal ends with the amniotic cavity. Later, both cranial and caudal neuropores close rendering a closed neural tube which will form the future spinal cord and brain vesicles [13, 51, 52].

Cells from the lateral border of the neuroectoderm form the ___neural crest___ (Fig. **4C**). The neural crest cells are multipotent migratory cells characteristic to the vertebrate embryo. These cells detach from the neighboring cells and migrate to the underlying mesoderm and undergo an epithelial-to-mesenchymal transition. The migrating crest cells give rise to a heterogeneous collection of structures as adrenal medulla, dorsal root ganglia and sympathetic chain ganglia. In addition, they contribute to formation of other tissues as pigment cells, smooth muscles and craniofacial cartilage and bone [53, 54].

Morphogenetic Changes of the Mesodermal Germ Layer

The ingressed cells that lie between the definitive endoderm and the epiblast form the intraembryonic mesoderm. Initially, these cells arrange on each side of the midline to form the *paraxial mesoderm*. More laterally, the mesodermal cells form the ___lateral plate mesoderm___. Many cavities appear and communicate inside the lateral plate mesoderm to form the ___intraembryonic cavity___ which communicates with the extraembryonic cavity. The wall of the cavity which is related to the amnion is called the parietal mesoderm while that in relation to the yolk sac is called visceral mesoderm. ___Intermediate mesoderm___ intervenes between paraxial mesoderm and lateral plate [13].

The paraxial mesoderm begins to be organised into segments termed ___somites___. These somites form periodically in an obvious head-to-tail sequence. Thus, the age of the embryo can be accurately determined during this early time by counting somites. However, the paraxial mesoderm in the future head region forms bands of cells that remain unsegmented as the head mesoderm [34]. Later, each somite is divided obliquely into ventromedial part called sclerotome and dorsolateral part called dermomyotome. Many structures are derived from somites including the vertebral column, part of the occipital bone, the voluntary musculature of neck, body wall and limbs and also the dermis of the back [34, 55, 56].

The cells of the intermediate mesoderm differentiate into segmental cell clusters in the cephlaic part of the embryo termed ___nephrotomes___ and caudal unsegmented mass of tissue termed ___nephrogenic cord___. Urogenital structures develop from this intermediate mesoderm [13, 34].

Morphogenetic Changes of the Endodermal Germ Layer

The first ingressing epiblast cells invade the hypoblast and displace its cells thus create the definitive endoderm. As a result of embryonic folding, the endodermal germ layer is incorporated inside the body of the embryo to form the primitive gut tube which is divided into *__foregut__*, *__midgut__*, and *__hind gut__*. The gut tube communicates with the yolk sac at the region of the midgut by the vitelline duct. The cephalic end of the foregut is closed by the oropharyngeal membrane which when ruptures the oral cavity connects with the primitive gut. Similarly, the hind gut is bounded caudally by the cloacal membrane which when ruptures form the opening for the anus. Definitive endoderm produces the epithelial lining of the alimentary tract, respiratory tract and urinary bladder. In addition, it contributes in formation of a variety of multiple cells and tissues including liver hepatocytes, pancreatic ß cells, lung alveolar cells, thymus and thyroid [57, 58].

EMBRYONIC STEM CELLS

Introduction

The study of early development of mammalian embryos suggests that a diverse range of pluripotent cells exists *in vivo*. As previously mentioned, the blastocyst stage shows two distinct lineages that possess morphological and molecular differences: the trophectoderm and the inner cell mass. Later, the ICM segregates into epiblast and hypoblast, which differ in their molecular signature [59, 60]. The epiblast cells are pluripotent and they differentiate during gastrulation, to give rise to the three primary germ layers from which all adult tissues are derived. After the process of gastrulation is completed, the only source of pluripotent stem cells in the embryo is the primordial germ cells derived from the epiblast.

__ES (Embryonic stem)__ cells are derived mostly from pre-implantation embryos and they represent permanent pluripotent stem cell lines. Currently, there is growing interest to study the potential therapeutic uses of human ES cells. ES cells originate from the inner cell mass of the blastocyst. Being pluripotent, these cells have the ability to give rise to different cell types of the body. After implantation, they generate other cell types with more restricted developmental potential. However, *in vitro* and under certain culture conditions, ICM cells possess the capacity to renew indefinitely in the form of ES cells which retain the pluripotency character. The first reports declaring the generation of mouse ES cells were in 1981 [61, 62], but derivation of human ES cells was first reported in 1998 [63]. The field of stem cell research is rapidly growing and new pluripotent cell lines are progressively isolated. Several research groups have succeeded in generation of stable pluripotent cells from rat and mouse epiplast of post-implantation embryos [64, 65].

Embryonic Sources of Stem Cell Lines

Currently, human pluripotent stem cell lines have been derived from many embryonic sources. In these approaches, viable cells are isolated during an early phase of development and then, they are grown in suitable culture conditions. One of the methods used for generation of human ES cells is the use of embryos that have been collected for ***IVF (in vitro fertilization)*** and are no longer needed for reproductive issues. The ICM cells are derived from a donated blastocyst at approximately five days after IVF and then placed into a specialized culture medium. Stem cell lines are generated and can be directed to give rise to cells and tissues of different lineages [66 - 68] (Fig. **5**). However, ethical considerations may restrict the use of this approach as the embryonic cells are derived from viable embryos. Moreover, cells derived by this approach may encounter immune rejection [13]. ES cell lines can be also generated from IVF embryos which are discarded from implantation in a woman seeking fertility treatment. Although these embryos had stopped dividing, the human stem cells derived from them are still pluripotent and have the ability to generate cells from different germ layers [69, 70].

Another method for production of human stem cell lines is the isolation of primordial germ cells from post implanted embryos (aged 5-7 weeks). Depending on the sex of the developing embryo, these primordial germ cells are directed to produce either oocytes or sperm cells after being placed into culture. Then, embryonic germ cell lines are generated which resemble the ES cells in many properties [71, 72]. However, the clinical use of these cells needs a comprehensive understanding of their generation and behavior. Most recently, a new approach called ***somatic nuclear transfer*** or ***therapeutic cloning*** is recorded. In this approach, nuclei are taken from adult cells as skin and reintroduced into enucleated oocytes. Stimulation of oocytes allows them to produce blastocysts, and then ES cells are isolated. Being derived from the host, these cells possess genetic compatibility and this technique gains more acceptability because no fertilization is involved [13].

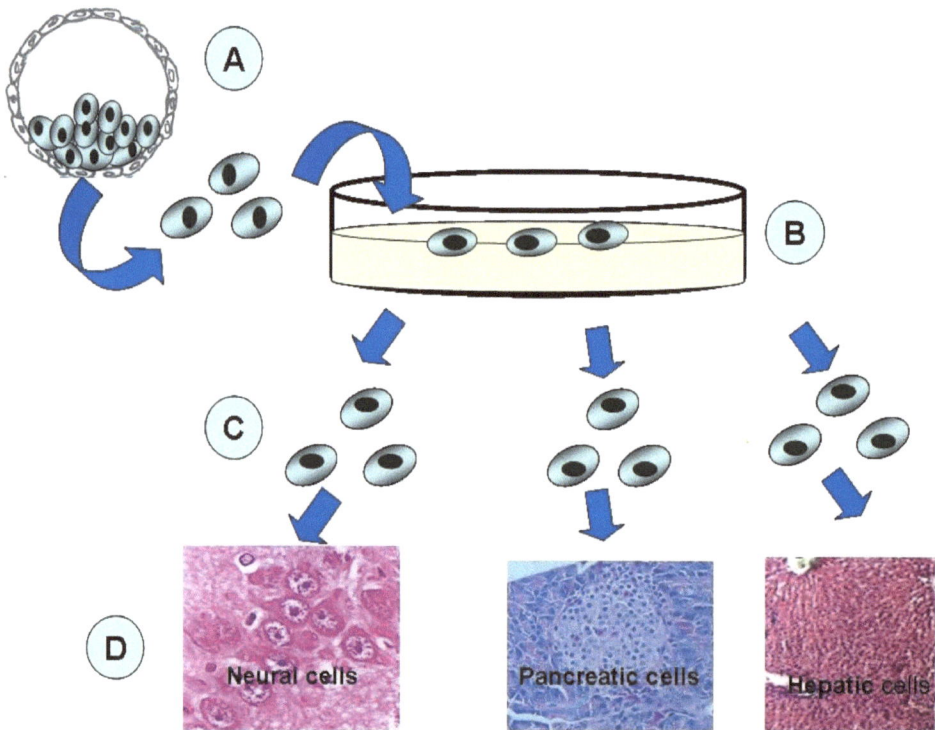

Fig. (5). Embryonic stem cells: **A**. Derivation of pluripotent cells from inner cell mass of the blastocyst. **B**. Incubation of derived cells in culture. **C**. The cells divide to make copies of themselves. **D**. The cells can differentiate into different types of cells from various lineages.

Non-Embryonic Sources of Stem Cell Lines

Stem cells can be also derived from non-embryonic tissues like placenta, amniotic fluid and umbilical cord. Other types of stem cells can be also obtained later in development or from the adult from particular tissues as hematopoietic or neural stem cells. Although these cells are expected to differentiate to the tissue from which they originate, great evidence shows their potential to produce other cell types under appropriate environmental conditions [73 - 76]. Recent researches have used adult stem cells extracted from rat brains in the treatment of Parkinson's disease in rats [13]. Thus, appropriate extrinsic environmental conditions can overcome stem cell restrictions and increase their plasticity. Currently, researches are done in many laboratories in order to achieve this aim.

Embryonic Stem Cell Generation and Fate

Some studies on mouse embryos have explained that ICM cells change significantly during derivation of ES cells [77]. During this phase, genes controlling the self renewal are up-regulated while genes controlling developmental differentiation are down-regulated. These changes are associated also with alterations of particular epigenetic regulators. Thus, the developmental program which occurs normally *in vivo* is arrested to allow the conversion of ICM cells to ES cells *in vitro*. Although there is acquisition of unlimited self renewal, the cells still retain the ability to differentiate into different cell types. Changes in epigenetic regulators render the acquired epigenotype stable which is important for the inherent plasticity of ES cells. Embryonic stem cells possess some new properties that are not found in cells of intact embryos. When ES cells exposed to tissue culture, they are stimulated by a variety of extrinsic factors that are not found *in vivo*. Adaptation of ES cells to these culture conditions allows them to gain new functions, and thus they possess the ability to proliferate indefinitely in an undifferentiated manner [78, 79].

Since 1998, many researches have been done to study the function of different genes that control the differentiation of human ES cells to produce different cell types. Actually, numerous factors influence the process of differentiation. Right knowledge of techniques enabling ES cells to produce different cell types can help scientists to use them in research work, transplantation therapy and drug discovery [80 - 82]. However, a more holistic understanding of the molecular mechanisms that govern the differentiation of pluripotent cells must be needed before the use of stem cells in clinical application. Extensive studies must be done to follow transplanted cells to be sure that they do not continue dividing which may lead to formation of a stem cell tumor. Moreover, scientists must be sure that transplanted cells will not be rejected by the patient's immune system.

Clinical Application of Embryonic Stem Cells

The pluripotency of ES cells offers an equivalent source for human tissues that can be used in basic research. For decades, researchers have relied on animal models for testing the efficacy and safety of drugs. However, because animal and human tissues possess many species- specific differences, many hazards have been encountered [83, 84]. Hence, cells derived from human ES cells can be used for testing the drugs to ensure more effective and safer treatments.

Human ES cells also can be used for treatment of some degenerative diseases by producing tissues for transplantation therapies. Cells derived from human ES cells may replace dopaminergic neurons in Parkinson's disease or insulin producing cells in diabetes, and thus could generate a lifelong treatment for multiple disorders [85 - 90]. Human embryonic stem cell lines may provide the scientists with a broad understanding of the differentiation of human tissues which may help them in managing infertility, birth defects, abortions, and intrauterine fetal deaths. Moreover, scientists can use stem cell lines derived from embryos identified with genetic disorders, such as thalessemia, muscular dystrophy, Marfan syndrome, neurofibromatosis and other genetically- based disorders for more understanding and exploration of these diseases [91].

CONFLICT OF INTEREST

The author confirms that author has no conflict of interest to declare for this publication.

ACKNOWLEDGEMENTS

Declared none.

REFERENCES

[1]   Saladin KS. Human development. In: Saladin KS, Ed. Anatomy and physiology (the unity of form and function). McGraw- Hill Companies. 2007; pp. 1106-37.

[2]   Schmitt S. From eggs to fossils: epigenesis and transformation of species in Pander's biology. Int J Dev Biol 2005; 49(1): 1-8.
[http://dx.doi.org/10.1387/ijdb.041908ss] [PMID: 15744661]

[3]   Sterba O. Staging and ageing of mammalian embryos and fetuses. Acta Vet Brno 1995; 64: 83-9.
[http://dx.doi.org/10.2754/avb199564010083]

[4]   O'Rahilly R, Müller F. Prenatal ages and stages-measures and errors. Teratology 2000; 61(5): 382-4.
[http://dx.doi.org/10.1002/(SICI)1096-9926(200005)61:5<382::AID-TERA10>3.0.CO;2-5]   [PMID: 10777834]

[5]   Berg DK, Smith CS, Pearton DJ, *et al.* Trophectoderm lineage determination in cattle. Dev Cell 2011; 20(2): 244-55.
[http://dx.doi.org/10.1016/j.devcel.2011.01.003] [PMID: 21316591]

[6]   Hall VJ, Christensen J, Gao Y, Schmidt MH, Hyttel P. Porcine pluripotency cell signaling develops from the inner cell mass to the epiblast during early development. Dev Dyn 2009; 238(8): 2014-24.
[http://dx.doi.org/10.1002/dvdy.22027] [PMID: 19618464]

[7]   Kirchhof N, Carnwath JW, Lemme E, Anastassiadis K, Schöler H, Niemann H. Expression pattern of Oct-4 in preimplantation embryos of different species. Biol Reprod 2000; 63(6): 1698-705.
[http://dx.doi.org/10.1095/biolreprod63.6.1698] [PMID: 11090438]

[8]   Kuijk EW, van Tol LT, Van de Velde H, *et al.* The roles of FGF and MAP kinase signaling in the segregation of the epiblast and hypoblast cell lineages in bovine and human embryos. Development 2012; 139(5): 871-82.
[http://dx.doi.org/10.1242/dev.071688] [PMID: 22278923]

[9]   Johnson MH, Selwood L. Nomenclature of early development in mammals. Reprod Fertil Dev 1996; 8(4): 759-64.
[http://dx.doi.org/10.1071/RD9960759] [PMID: 8870096]

[10]   Ikawa M, Wada I, Kominami K, *et al.* The putative chaperone calmegin is required for sperm fertility. Nature 1997; 387(6633): 607-11.
[http://dx.doi.org/10.1038/42484] [PMID: 9177349]

[11]   Ikawa M, Inoue N, Benham AM, Okabe M. Fertilization: a sperm's journey to and interaction with the oocyte. J Clin Invest 2010; 120(4): 984-94.
[http://dx.doi.org/10.1172/JCI41585] [PMID: 20364096]

[12]   Nakanishi T, Isotani A, Yamaguchi R, *et al.* Selective passage through the uterotubal junction of sperm from a mixed population produced by chimeras of calmegin-knockout and wild-type male mice. Biol Reprod 2004; 71(3): 959-65.
[http://dx.doi.org/10.1095/biolreprod.104.028647] [PMID: 15151931]

[13]   Sadler TW. ED Langman's medical embryology. 12th ed. Philadelphia: Wolters Kluwer Health/Lippincott Williams & Wilkins 2012; pp. 29-85.

[14]   Yanagimachi R. Mammalian fertilization. In: Knobil E, Neill ID, Eds. Physiology of Reproduction. 2nd ed. New York, NY: Raven Press 1994; pp. 189-317.

[15]   Tulsiani DR, Abou-Haila A, Loeser CR, Pereira BM. The biological and functional significance of the sperm acrosome and acrosomal enzymes in mammalian fertilization. Exp Cell Res 1998; 240(2): 151-

64.
[http://dx.doi.org/10.1006/excr.1998.3943] [PMID: 9596988]

[16]   Bailey JL. Factors regulating sperm capacitation. Syst Biol Reprod Med 2010; 56(5): 334-48.
[http://dx.doi.org/10.3109/19396368.2010.512377] [PMID: 20849222]

[17]   Visconti PE, Krapf D, de la Vega-Beltrán JL, Acevedo JJ, Darszon A. Ion channels, phosphorylation and mammalian sperm capacitation. Asian J Androl 2011; 13(3): 395-405.
[http://dx.doi.org/10.1038/aja.2010.69] [PMID: 21540868]

[18]   Gahlay G, Gauthier L, Baibakov B, Epifano O, Dean J. Gamete recognition in mice depends on the cleavage status of an egg's zona pellucida protein. Science 2010; 329(5988): 216-9.
[http://dx.doi.org/10.1126/science.1188178] [PMID: 20616279]

[19]   Abou-haila A, Tulsiani DR. Signal transduction pathways that regulate sperm capacitation and the acrosome reaction. Arch Biochem Biophys 2009; 485(1): 72-81.
[http://dx.doi.org/10.1016/j.abb.2009.02.003] [PMID: 19217882]

[20]   Florman HM, Ducibella T. Fertilization in Mammals. In: Neill JD, Ed. Physiology of reproduction. 3rd ed. San Diego: Elsevier 2006; pp. 55-112.

[21]   Miyazaki S, Ito M. Calcium signals for egg activation in mammals. J Pharmacol Sci 2006; 100(5): 545-52.
[http://dx.doi.org/10.1254/jphs.CPJ06003X] [PMID: 16799264]

[22]   Burkart AD, Xiong B, Baibakov B, Jiménez-Movilla M, Dean J. Ovastacin, a cortical granule protease, cleaves ZP2 in the zona pellucida to prevent polyspermy. J Cell Biol 2012; 197(1): 37-44.
[http://dx.doi.org/10.1083/jcb.201112094] [PMID: 22472438]

[23]   Hamatani T, Carter MG, Sharov AA, Ko MS. Dynamics of global gene expression changes during mouse preimplantation development. Dev Cell 2004; 6(1): 117-31.
[http://dx.doi.org/10.1016/S1534-5807(03)00373-3] [PMID: 14723852]

[24]   Guo G, Huss M, Tong GQ, *et al.* Resolution of cell fate decisions revealed by single-cell gene expression analysis from zygote to blastocyst. Dev Cell 2010; 18(4): 675-85.
[http://dx.doi.org/10.1016/j.devcel.2010.02.012] [PMID: 20412781]

[25]   Walser CB, Lipshitz HD. Transcript clearance during the maternal-to-zygotic transition. Curr Opin Genet Dev 2011; 21(4): 431-43.
[http://dx.doi.org/10.1016/j.gde.2011.03.003] [PMID: 21497081]

[26]   Perez-Moreno M, Jamora C, Fuchs E. Sticky business: orchestrating cellular signals at adherens junctions. Cell 2003; 112(4): 535-48.
[http://dx.doi.org/10.1016/S0092-8674(03)00108-9] [PMID: 12600316]

[27]   Riethmacher D, Brinkmann V, Birchmeier C. A targeted mutation in the mouse E-cadherin gene results in defective preimplantation development Proc Natl cad Sci 1995; 92: 855-9.

[28]   Suzuki A, Ohno S. The PAR-aPKC system: lessons in polarity. J Cell Sci 2006; 119(Pt 6): 979-87.
[http://dx.doi.org/10.1242/jcs.02898] [PMID: 16525119]

[29]   Alarcon VB. Cell polarity regulator PARD6B is essential for trophectoderm formation in the preimplantation mouse embryo. Biol Reprod 2010; 83(3): 347-58.
[http://dx.doi.org/10.1095/biolreprod.110.084400] [PMID: 20505164]

[30]   Kidder BL, Palmer S. Examination of transcriptional networks reveals an important role for TCFAP2C, SMARCA4, and EOMES in trophoblast stem cell maintenance. Genome Res 2010; 20(4): 458-72.
[http://dx.doi.org/10.1101/gr.101469.109] [PMID: 20176728]

[31]   Yamanaka Y, Ralston A, Stephenson RO, Rossant J. Cell and molecular regulation of the mouse blastocyst. Dev Dyn 2006; 235(9): 2301-14.
[http://dx.doi.org/10.1002/dvdy.20844] [PMID: 16773657]

[32]   Barcroft LC, Offenberg H, Thomsen P, Watson AJ. Aquaporin proteins in murine trophectoderm mediate transepithelial water movements during cavitation. Dev Biol 2003; 256(2): 342-54.
[http://dx.doi.org/10.1016/S0012-1606(02)00127-6] [PMID: 12679107]

[33]   Standring S, Borley NR, Collins P, *et al.* EDs Gray's Anatomy: The anatomical basis of clinical practice. 40th ed. Churchill livingstone, Elsevier 2008; pp. 167-203.

[34]   Schoenwolf GC, Bleyl SB, Brauer PR, Francis-West PH. EDs Larsen's human embryology. 4th ed. Churchill livingstone, Elsevier 2009; pp. 39-131.

[35]   Rossant J, Tam PP. Blastocyst lineage formation, early embryonic asymmetries and axis patterning in the mouse. Development 2009; 136(5): 701-13.
[http://dx.doi.org/10.1242/dev.017178] [PMID: 19201946]

[36]   Beddington RS, Robertson EJ. Axis development and early asymmetry in mammals. Cell 1999; 96(2): 195-209.
[http://dx.doi.org/10.1016/S0092-8674(00)80560-7] [PMID: 9988215]

[37]   Wilcox AJ, Baird DD, Weinberg CR. Time of implantation of the conceptus and loss of pregnancy. N Engl J Med 1999; 340(23): 1796-9.
[http://dx.doi.org/10.1056/NEJM199906103402304] [PMID: 10362823]

[38]   Carson DD, Bagchi I, Dey SK, *et al.* Embryo implantation. Dev Biol 2000; 223(2): 217-37.
[http://dx.doi.org/10.1006/dbio.2000.9767] [PMID: 10882512]

[39]   Gray CA, Bartol FF, Tarleton BJ, *et al.* Developmental biology of uterine glands. Biol Reprod 2001; 65(5): 1311-23.
[http://dx.doi.org/10.1095/biolreprod65.5.1311] [PMID: 11673245]

[40]   Hoffman LH, Olson GE, Carson DD, Chilton BS. Progesterone and implanting blastocysts regulate Muc1 expression in rabbit uterine epithelium. Endocrinology 1998; 139(1): 266-71.
[PMID: 9421424]

[41]   Wimsatt WA. Some comparative aspects of implantation. Biol Reprod 1975; 12(1): 1-40.
[http://dx.doi.org/10.1095/biolreprod12.1.1] [PMID: 806310]

[42]   Dey SK. Visualizing Early Embryo Implantation Sites by Dye Injection. In: Nagy A, Ed. Manipulating the Mouse Embryo. Cold Spring Harbor, NY: Cold Spring Harbor Press 2003; pp. 10-1.

[43]   Enders AC. Trophoblast-uterine interactions in the first days of implantation: models for the study of implantation events in the human. Semin Reprod Med 2000; 18(3): 255-63.
[http://dx.doi.org/10.1055/s-2000-12563] [PMID: 11299964]

[44]   Tam PP, Behringer RR. Mouse gastrulation: the formation of a mammalian body plan. Mech Dev

1997; 68(1-2): 3-25.
[http://dx.doi.org/10.1016/S0925-4773(97)00123-8] [PMID: 9431800]

[45]   Tam PP. Postimplantation mouse development: whole embryo culture and micro-manipulation. Int J Dev Biol 1998; 42(7): 895-902.
[PMID: 9853819]

[46]   Williams M, Burdsal C, Periasamy A, Lewandoski M, Sutherland A. Mouse primitive streak forms in situ by initiation of epithelial to mesenchymal transition without migration of a cell population. Dev Dyn 2012; 241(2): 270-83.
[http://dx.doi.org/10.1002/dvdy.23711] [PMID: 22170865]

[47]   Kwon GS, Viotti M, Hadjantonakis AK. The endoderm of the mouse embryo arises by dynamic widespread intercalation of embryonic and extraembryonic lineages. Dev Cell 2008; 15(4): 509-20.
[http://dx.doi.org/10.1016/j.devcel.2008.07.017] [PMID: 18854136]

[48]   Müller F, O'Rahilly R. The prechordal plate, the rostral end of the notochord and nearby median features in staged human embryos. Cells Tissues Organs (Print) 2003; 173(1): 1-20.
[http://dx.doi.org/10.1159/000068214] [PMID: 12566624]

[49]   Yamanaka Y, Tamplin OJ, Beckers A, Gossler A, Rossant J. Live imaging and genetic analysis of mouse notochord formation reveals regional morphogenetic mechanisms. Dev Cell 2007; 13(6): 884-96.
[http://dx.doi.org/10.1016/j.devcel.2007.10.016] [PMID: 18061569]

[50]   Waterston RH, Lindblad-Toh K, Birney E, *et al.* Initial sequencing and comparative analysis of the mouse genome. Nature 2002; 420(6915): 520-62.
[http://dx.doi.org/10.1038/nature01262] [PMID: 12466850]

[51]   Colas JF, Schoenwolf GC. Towards a cellular and molecular understanding of neurulation. Dev Dyn 2001; 221(2): 117-45.
[http://dx.doi.org/10.1002/dvdy.1144] [PMID: 11376482]

[52]   Lowery LA, Sive H. Strategies of vertebrate neurulation and a re-evaluation of teleost neural tube formation. Mech Dev 2004; 121(10): 1189-97.
[http://dx.doi.org/10.1016/j.mod.2004.04.022] [PMID: 15327780]

[53]   Huang X, Saint-Jeannet JP. Induction of the neural crest and the opportunities of life on the edge. Dev Biol 2004; 275(1): 1-11.
[http://dx.doi.org/10.1016/j.ydbio.2004.07.033] [PMID: 15464568]

[54]   Stuhlmiller TJ, García-Castro MI. Current perspectives of the signaling pathways directing neural crest induction. Cell Mol Life Sci 2012; 69(22): 3715-37.
[http://dx.doi.org/10.1007/s00018-012-0991-8] [PMID: 22547091]

[55]   Dubrulle J, Pourquié O. Coupling segmentation to axis formation. Development 2004; 131(23): 5783-93.
[http://dx.doi.org/10.1242/dev.01519] [PMID: 15539483]

[56]   Mallo M, Vinagre T, Carapuço M. The road to the vertebral formula. Int J Dev Biol 2009; 53(8-10): 1469-81.
[http://dx.doi.org/10.1387/ijdb.072276mm] [PMID: 19247958]

[57]   Lu CC, Brennan J, Robertson EJ. From fertilization to gastrulation: axis formation in the mouse embryo. Curr Opin Genet Dev 2001; 11(4): 384-92.
[http://dx.doi.org/10.1016/S0959-437X(00)00208-2] [PMID: 11448624]

[58]   Zorn AM, Wells JM. Vertebrate endoderm development and organ formation. Annu Rev Cell Dev Biol 2009; 25: 221-51.
[http://dx.doi.org/10.1146/annurev.cellbio.042308.113344] [PMID: 19575677]

[59]   Chazaud C, Yamanaka Y, Pawson T, Rossant J. Early lineage segregation between epiblast and primitive endoderm in mouse blastocysts through the Grb2-MAPK pathway. Dev Cell 2006; 10(5): 615-24.
[http://dx.doi.org/10.1016/j.devcel.2006.02.020] [PMID: 16678776]

[60]   Plusa B, Piliszek A, Frankenberg S, Artus J, Hadjantonakis AK. Distinct sequential cell behaviours direct primitive endoderm formation in the mouse blastocyst. Development 2008; 135(18): 3081-91.
[http://dx.doi.org/10.1242/dev.021519] [PMID: 18725515]

[61]   Evans MJ, Kaufman MH. Establishment in culture of pluripotential cells from mouse embryos. Nature 1981; 292(5819): 154-6.
[http://dx.doi.org/10.1038/292154a0] [PMID: 7242681]

[62]   Martin GR. Isolation of a pluripotent cell line from early mouse embryos cultured in medium conditioned by teratocarcinoma stem cells. Proc Natl Acad Sci USA 1981; 78(12): 7634-8.
[http://dx.doi.org/10.1073/pnas.78.12.7634] [PMID: 6950406]

[63]   Thomson JA, Itskovitz-Eldor J, Shapiro SS, *et al.* Embryonic stem cell lines derived from human blastocysts. Science 1998; 282(5391): 1145-7.
[http://dx.doi.org/10.1126/science.282.5391.1145] [PMID: 9804556]

[64]   Tesar PJ, Chenoweth JG, Brook FA, *et al.* New cell lines from mouse epiblast share defining features with human embryonic stem cells. Nature 2007; 448(7150): 196-9.
[http://dx.doi.org/10.1038/nature05972] [PMID: 17597760]

[65]   Brons IG, Smithers LE, Trotter MW, *et al.* Derivation of pluripotent epiblast stem cells from mammalian embryos. Nature 2007; 448(7150): 191-5.
[http://dx.doi.org/10.1038/nature05950] [PMID: 17597762]

[66]   He JQ, Ma Y, Lee Y, Thomson JA, Kamp TJ. Human embryonic stem cells develop into multiple types of cardiac myocytes: action potential characterization. Circ Res 2003; 93(1): 32-9.
[http://dx.doi.org/10.1161/01.RES.0000080317.92718.99] [PMID: 12791707]

[67]   Zhang SC, Wernig M, Duncan ID, Brüstle O, Thomson JA. *In vitro* differentiation of transplantable neural precursors from human embryonic stem cells. Nat Biotechnol 2001; 19(12): 1129-33.
[http://dx.doi.org/10.1038/nbt1201-1129] [PMID: 11731781]

[68]   Wang L, Li L, Menendez P, Cerdan C, Bhatia M. Human embryonic stem cells maintained in the absence of mouse embryonic fibroblasts or conditioned media are capable of hematopoietic development. Blood 2005; 105(12): 4598-603.
[http://dx.doi.org/10.1182/blood-2004-10-4065] [PMID: 15718421]

[69]   Zhang X, Stojkovic P, Przyborski S, *et al.* Derivation of human embryonic stem cells from developing and arrested embryos. Stem Cells 2006; 24(12): 2669-76.

[http://dx.doi.org/10.1634/stemcells.2006-0377] [PMID: 16990582]

[70]   Landry DW, Zucker HA. Embryonic death and the creation of human embryonic stem cells. J Clin Invest 2004; 114(9): 1184-6.
[http://dx.doi.org/10.1172/JCI23065] [PMID: 15520846]

[71]   Turnpenny L, Brickwood S, Spalluto CM, *et al.* Derivation of human embryonic germ cells: an alternative source of pluripotent stem cells. Stem Cells 2003; 21(5): 598-609.
[http://dx.doi.org/10.1634/stemcells.21-5-598] [PMID: 12968114]

[72]   Aflatoonian B, Moore H. Human primordial germ cells and embryonic germ cells, and their use in cell therapy. Curr Opin Biotechnol 2005; 16(5): 530-5.
[http://dx.doi.org/10.1016/j.copbio.2005.08.008] [PMID: 16154336]

[73]   Jackson KA, Majka SM, Wang H, *et al.* Regeneration of ischemic cardiac muscle and vascular endothelium by adult stem cells. J Clin Invest 2001; 107(11): 1395-402.
[http://dx.doi.org/10.1172/JCI12150] [PMID: 11390421]

[74]   Lagasse E, Connors H, Al-Dhalimy M, *et al.* Purified hematopoietic stem cells can differentiate into hepatocytes *in vivo.* Nat Med 2000; 6(11): 1229-34.
[http://dx.doi.org/10.1038/81326] [PMID: 11062533]

[75]   Krause DS, Theise ND, Collector MI, *et al.* Multi-organ, multi-lineage engraftment by a single bone marrow-derived stem cell. Cell 2001; 105(3): 369-77.
[http://dx.doi.org/10.1016/S0092-8674(01)00328-2] [PMID: 11348593]

[76]   Galli R, Borello U, Gritti A, *et al.* Skeletal myogenic potential of human and mouse neural stem cells. Nat Neurosci 2000; 3(10): 986-91.
[http://dx.doi.org/10.1038/79924] [PMID: 11017170]

[77]   Tang F, Barbacioru C, Bao S, *et al.* Tracing the derivation of embryonic stem cells from the inner cell mass by single-cell RNA-Seq analysis. Cell Stem Cell 2010; 6(5): 468-78.
[http://dx.doi.org/10.1016/j.stem.2010.03.015] [PMID: 20452321]

[78]   Buehr M, Smith A. Genesis of embryonic stem cells. Philos Trans R Soc Lond B Biol Sci 2003; 358(1436): 1397-402.
[http://dx.doi.org/10.1098/rstb.2003.1327] [PMID: 14511487]

[79]   Rossant J. Stem cells from the Mammalian blastocyst. Stem Cells 2001; 19(6): 477-82.
[http://dx.doi.org/10.1634/stemcells.19-6-477] [PMID: 11713338]

[80]   Eiges R, Schuldiner M, Drukker M, Yanuka O, Itskovitz-Eldor J, Benvenisty N. Establishment of human embryonic stem cell-transfected clones carrying a marker for undifferentiated cells. Curr Biol 2001; 11(7): 514-8.
[http://dx.doi.org/10.1016/S0960-9822(01)00144-0] [PMID: 11413002]

[81]   Gropp M, Itsykson P, Singer O, *et al.* Stable genetic modification of human embryonic stem cells by lentiviral vectors. Mol Ther 2003; 7(2): 281-7.
[http://dx.doi.org/10.1016/S1525-0016(02)00047-3] [PMID: 12597917]

[82]   Lakshmipathy U, Pelacho B, Sudo K, *et al.* Efficient transfection of embryonic and adult stem cells. Stem Cells 2004; 22(4): 531-43.
[http://dx.doi.org/10.1634/stemcells.22-4-531] [PMID: 15277699]

[83]   Bremer S, Hartung T. The use of embryonic stem cells for regulatory developmental toxicity testing *in vitro*--the current status of test development. Curr Pharm Des 2004; 10(22): 2733-47.
[http://dx.doi.org/10.2174/1381612043383700] [PMID: 15320739]

[84]   Rolletschek A, Blyszczuk P, Wobus AM. Embryonic stem cell-derived cardiac, neuronal and pancreatic cells as model systems to study toxicological effects. Toxicol Lett 2004; 149(1-3): 361-9.
[http://dx.doi.org/10.1016/j.toxlet.2003.12.064] [PMID: 15093282]

[85]   Chiba S, Ikeda R, Kurokawa MS, *et al.* Anatomical and functional recovery by embryonic stem cell-derived neural tissue of a mouse model of brain damage. J Neurol Sci 2004; 219(1-2): 107-17.
[http://dx.doi.org/10.1016/j.jns.2004.01.006] [PMID: 15050446]

[86]   Kim D, Gu Y, Ishii M, *et al.* In vivo functioning and transplantable mature pancreatic islet-like cell clusters differentiated from embryonic stem cell. Pancreas 2003; 27(2): e34-41.
[http://dx.doi.org/10.1097/00006676-200308000-00021] [PMID: 12883277]

[87]   Lee DH, Park S, Kim EY, *et al.* Enhancement of re-closure capacity by the intra-amniotic injection of human embryonic stem cells in surgically induced spinal open neural tube defects in chick embryos. Neurosci Lett 2004; 364(2): 98-100.
[http://dx.doi.org/10.1016/j.neulet.2004.04.033] [PMID: 15196686]

[88]   Min JY, Yang Y, Converso KL, *et al.* Transplantation of embryonic stem cells improves cardiac function in postinfarcted rats. J Appl Physiol 2002; 92(1): 288-96.
[PMID: 11744672]

[89]   Park S, Kim EY, Ghil GS, *et al.* Genetically modified human embryonic stem cells relieve symptomatic motor behavior in a rat model of Parkinson's disease. Neurosci Lett 2003; 353(2): 91-4.
[http://dx.doi.org/10.1016/j.neulet.2003.08.082] [PMID: 14664908]

[90]   von Unge M, Dirckx JJ, Olivius NP. Embryonic stem cells enhance the healing of tympanic membrane perforations. Int J Pediatr Otorhinolaryngol 2003; 67(3): 215-9.
[http://dx.doi.org/10.1016/S0165-5876(02)00371-3] [PMID: 12633919]

[91]   Verlinsky Y, Strelchenko N, Kukharenko V, *et al.* Human embryonic stem cell lines with genetic disorders. Reprod Biomed Online 2005; 10(1): 105-10.
[http://dx.doi.org/10.1016/S1472-6483(10)60810-3] [PMID: 15705304]

CHAPTER 3

Gene and Signals Regulating Stem Cell Fates

Mohamed Berika[1,2,§], **John Ku**[2,§], **Haifen Huang**[3,§] and **Ahmed El-Hashash**[3,*]

[1] *Anatomy Department, Faculty of Medicine, Mansoura University, Mansoura, Egypt*

[2] *Rehabilitation Science Department, College of Applied Medical Sciences, King Saud University, KSA*

[3] *Stem Cells, Regenerative Medicine and Developmental Biology Program, Childrens Hospital Los Angeles, Keck School of Medicine and Ostrow School of Dentistry, University of Southern California, 4661 Sunset Boulevard, Los Angeles, CA 90027, USA*

Abstract: Recent developments have revealed more about the signaling pathways governing hematopoietic stem cell (HSC) fate; several developmentally conserved pathways were identified, including Notch, Smad pathways, Sonic hedgehog (Shh), and Wingless-type (Wnt). These findings contribute to our understanding of HSC regulation and provide information of interactions within the bone marrow environment where HSCs reside. The signaling pathways that contribute to HSC regulation are further discussed in this chapter.

Keywords: Asymmetric divisions, Cell fate, Hematopoietic cells, Signaling pathway, Symmetric.

INTRODUCTION

Maintaining lifelong organ function requires a continual replacement of cells. This is accomplished through resident stem cell populations, and illustrated particularly in tissues with high turnover rates. Formation of all blood cell lineages is the responsibility of stem cells residing in the adult bone marrow.

* **Correspondence author Ahmed El-Hashash:** Developmental Biology, Regenerative Medicine and Stem Cell Program, Saban Research Institute, Childrens Hospital Los Angeles, 4661 Sunset Boulevard MS 35, Los Angeles, California 90027, USA; Tel: 323-361-2764, 323-361-2258; Fax: 323-361-3613; E-mail: aelhashash@chla.usc.edu
§ Equal first author

These HSCs need to be capable of both differentiation and self- renewal to maintain blood components and stabilize the current HSC population within the bone marrow. Self-renewing division occurs by either a symmetric or asymmetric process. A symmetric division occurs when stem cells undergo self-renewal division producing two daughter cells retaining the stem cell properties. Whereas, an asymmetric self-renewing division occurs when resultant two daughter cells have different fates, as only one cell maintains the stem cell properties. The former process may expand the stem-cell pool, as both daughter cells become stem cells; this symmetric process may be useful in healing hematopoietic injury or after transplantation. While, the asymmetric process allows for a steady population of HSCs. Self-renewal is an important aspect for lifelong hematopoiesis.

THE HIERARCHY OF HEMATOPOIETIC CELLS

HSCs give a progeny that progressively loses self- renewing capacity. This allows for differentiation into functionally mature cells. Functional HSCs lack expression of several surface markers which are normally found on mature blood cells. High levels of Sca1 and c-kit are processed by HSCs. Sca1 and c-kit are known as the LSK compartment (lineage-/Sca1+/c-kit+)[1, 2]. All blood cell lineages were restored by several groups after transplanting a single HSC into lethally irradiated mice [3, 4]. It has also been shown that 100 LSK cells are sufficient for long-term re-population of blood cell lineages in lethally irradiated mice [2].

There are two types of reconstituting HSCs: long-term and short-term HSCs. Long-term HSCs (LT -HSCs) can sustain lifelong hematopoiesis by possessing a large capacity for self-renewal. Short-term HSCs (ST-HSCs) are limited in self-renewal and generate blood cells for a finite amount of time *in vivo*. The LSK compartment contains both LT-HSCs, ST-HSCs in addition to multipotent progenitor cells (MPPs) [5 - 7].

Isolation of human HSCs is difficult due to the lack of appropriate *in vivo* assays. For example, non-obese diabetic/severe combined immune deficient (NOD/SCID) mouse strains generally have low engraftment of human cells and do not live long enough for long-term analysis. Some have chosen to perform serial transplantations in NOD/SCID assays, however the HSCs in this case are kept under constant stress and the data may be unreliable.

REGULATION OF HEMATOPOIESIS

Both intrinsic and extrinsic signals regulate hematopoiesis to accommodate for a variety of situations such as acute blood loss or infection [8 - 10]. Many factors were identified in HSC regulation such as growth factors, transcription factors, cytokines, cell-cycle regulators, and chromatin modifiers. Discovering the roles of these factors in signaling pathways is of utmost importance. During development, many signaling pathways are responsible in determining embryonic structure and cell fate. It is probable that somatic stem-cell compartments of adults are influenced by developmentally conserved signals.

CYTOKINES: A KEY REGULATOR OF HSCS

Many attempts have been made in using classic hematopoietic cytokines to expand HSCs *in vitro*. However, many have failed in preventing HSCs from differentiating. One study observed a re-population of murine HSCs in a 10-day culture protocol containing interleukin (IL)-11, Flt-3 ligand, and stem-cell factor (SCF) [11]. However, other studies have shown that these conditions failed to expand HSCs derived from fetal liver. They found that Flt-3 receptor is not expressed on LT-HSCs. In addition, they noticed a possible role of the gp130 protein in HSC self-renewal as it constitutes part of the receptor for IL-6 and IL-11. Furthermore, elimination of the IL-11 receptor does not affect hematopoiesis [6, 12 - 14]. These findings make for a lot of uncertainty. Studies with purified murine HSCs proved that the receptors for thrombopoietin (TPO) and SCF are both expressed on repopulating HSCs [5, 6, 15 - 18].

Genetic mutations in TPO or its receptor, c-mpl, have shown a reduction in HSCs in mice [15, 16]. Reports have indicated that TPO supports viability and suppresses apoptosis of HSCs [19, 20]. Altogether, TPO may counteract apoptosis rather than promoting HSC expansion [18].

The intracellular adaptor molecule Lnk, acts as a broad inhibitor of cytokine signaling pathways, including SCF, TPO, IL-3, IL-7 and erythropoietin, [18, 21 -

24]. Lnk deficient Mice reportedly had increased HSC numbers and an elevated re-population function [18, 21, 25]. Lnk-/- HSCs seem to be dependent on TPO and thus the function of Lnk is to act as an inhibitor of TPO signaling in HSCs. TPO and SCF are the most important positive HSC regulators to date.

FUNCTIONS OF NOTCH PATHWAY IN HSCS

In HSCs: Notch is a known oncogene that plays a role in lineage specification of lymphopoiesis, by driving differentiation toward T cells [26]. Notch and Notch-ligand mRNA are both expressed in primitive hematopoietic cells and those in the putative HSC microenvironment [27, 28]. *In vitro* the Notch ligand families, Delta and Jagged, have been shown to expand both human and murine hematopoietic progenitor cells [28 - 34]. Primitive hematopoietic progenitor cells were immortalized by an enforced activation of the Notch signaling pathway or the downstream target Hes-1, [35 - 37]. Calvi *et al.* showed how a constitutively active parathyroid hormone receptor expanded osteoblasts by stimulating expression of Jagged-1. These osteoblasts increased the self-renewal capacity of primitive hematopoietic cells. Increased levels of Notch intracellular domain (NICD) were noticed in LSK cells from these type of mice. This study shows osteoblasts as part of the HSC niche and also the importance of the Jagged/Notch signaling activation in extracellular regulation of HSC self-renewal [38]. HSCs deficient in Notch-1 were shown to reconstitute Jagged-1 deficient hosts, possibly through a compensatory mechanism with other Notch receptors and ligands [39]. Elimination of all Notch signaling pathways would help in figuring out *ex vivo* expansion of HSCs in the future.

FUNCTIONS OF WNT SIGNALING PATHWAY IN THE SELF-RENEWAL OF HSCS

Wnt proteins have been identified in both fetal and adult bone marrow. Utilizing Wnt proteins in controlled reconstitution assays has demonstrated an up-regulation in the self-renewal process for both murine and human HSCs *in vitro* [40]. The end product of Wnt signaling is the transcription of target genes that are influenced by the expression of β-catenin. The expression of β-catenin *in vivo*, however, exhibits a phenotype that contradicts its *in vitro* counterpart [41 - 43].

This difference upholds the complexity of the *in vivo* environment, suggesting backup or alternative mechanisms that counter the impairment of β-catenin in the signaling pathway [41]. The exact mechanism by which the Wnt pathway operates is yet to be determined.

INTERACTION BETWEEN NOTCH AND WNT SIGNAL PATHWAYS IN HSC

The Wnt and Notch signaling pathways are only two out of many other pathways that regulate the development of HSCs. More specifically, Wnt signaling appears to promote the proliferation of the LSK compartment while the Notch pathway ensures that these cells remain undifferentiated [44]. The two pathways are co-dependent as steps in each pathway tend to overlap, in which respective ligands influence and interact with ligands belonging of the other pathway [45].

FUNCTIONS OF SMAD SIGNALING PATHWAY

The Smad pathway consists of multiple signal transduction pathways that often overlap. It is for this reason that the Smad pathway has yet to be constructed and fully understood. These pathways may diverge and converge, further elaborating the Smad pathway. Fatality associated with knockout mice models also renders the study of the Smad pathway difficult. The Smad pathway itself is a negative regulator of HSC growth *in vivo*. Ligands that play a central role in mediating the proliferative and self-renewal capabilities of HSCs include the transforming growth factor- β (TGF-β) and bone morphogenetic proteins (BMPs). *In vitro* studies have demonstrated that TGF-β severely inhibits the growth of HSCs while BMPs maintains their rate of self-renewal [46 - 48]. The same studies conducted *in vivo*, however, have resulted in contradictory results with a majority of them lacking the expected *in vitro* phenotype. These inconsistencies suggest alternative mechanisms, generating redundancy in the Smad pathway [49, 50]. In an effort to overcome these often conflicting results, researchers conducted *in vivo* studies and discovered additional ligands that arbitrate signals in the Smad pathway such as Smad7 and Smad4. Smad7 and Smad4 demonstrate opposing regulatory function. Smad7 positively regulates the self-renewal process while Smad4 exhibits a negative effect on populations of HSCs [51 - 53]. The exact mechanisms through which these ligands interact with other intermediaries and with each other have still not been clearly defined. Different combinations of ligands and the extent of each interaction have had entirely different effects on the HSC compartment.

FUNCTIONAL ROLES OF ANGIOPOIETIN-LIKE PROTEINS IN HSCS

During early stages of development, HSC growth occurs substantially in the fetal liver (FL). Angiopoietin-like (Angptl) proteins are growth factors that amplify HSCs of the FL. The behavior of these Angptl proteins is not yet fully understood and the mechanisms by which they expand HSCs are still being investigated. However, there is an evidence suggesting that Angptl proteins interact with HSCs directly.

HSC NICHE

Two types of connective tissue line the bone. The endosteum lines the inner cavity or the bone marrow while the periosteum lines the outer layer of bone. It is on the endosteal surface of bone where an HSC niche is established. The HSC niche is composed of the microenvironment in the BM where HSCs develop and expand [54]. Osteoblasts, cells that form bone tissue, are important components of this niche. A lack of osteoblasts *in vivo* has resulted in decreased cellularity, further supporting the notion that osteoblasts are necessary in this microenvironment [55]. Other constituents that affect the HSC niche include soluble factors, matrix proteins, and cellular ligands [56, 57]. Furthermore, calcium gradients are necessary for maintaining localization of HSCs in their respective regions in the bone marrow, thus cultivating the HSC niche [58]. Environmental stress can have detrimental effects on the HSC niche. A distinctive feature of the HSC niche is that it is constantly in a state of hypoxia [59]. Oxidative stress can result in early BM failure [60]. More specifically, oxygen levels must be kept low or it may increase the frequency of HSC cycling and thereby exhausting its functionality at a faster rate [61]. The HSC niche thus needs to maintain a hypoxic microenvironment in order to function properly. Before HSCs establish themselves in the bone marrow, however, they start off in the fetal liver (FL). Primitive HSCs cycle exponentially in FL but recent studies in mice have proposed an intrinsic mechanism by which HSCs transform from this state to that of quiescence in 3 to 4 weeks' time [62]. The HSC niche is a complex system regulated and affected by both intrinsic and extrinsic factors that are still being explored at various levels of stem cell research.

CLINICAL SIGNIFICANCES AND APPLICATIONS OF HSCS IN CELL AND GENE THERAPY

As it stands, umbilical cord blood is commonly being used as a reliable source of HSCs due to a variety and diversity of cells that exist in CB banks [63]. Since CB samples are limited, researchers have developed way to amplify HSCs of interest *ex vivo* for both research and medicinal purposes.

In order to successfully maintain a constant population of HSCs for study and use, self-renewal must be symmetric [64]. HSCs, unfortunately, divide asymmetrically *in vitro*. As a result, methods to expand CB stem cells rely on regulators that promote self-renewal and processes that inhibit inactivation, differentiation, and apoptosis. The transfer of appropriate soluble factors is often the preferred approach to stimulating expansion. Their expression in HSCs of interest is safer compared to, for instance, activation mediated by a viral gene transfer. It has already been determined that HoxB4 and Angptl proteins can readily expand murine HSCs [65 - 67]. Activation of pathways such as Notch and Wnt may also amplify self-renewal properties [28]. It is important to note that most of these advances in stem cell knowledge have resulted from utilizing murine models. Their human counterparts may be similar or significantly different with respects to the nature of existing ligands, self-renewal capabilities, and properties of proliferation. Results from *in vitro* studies may also contradict with results from the same studies conducted *in vivo*. Mechanisms that explain the behavior of stem cell constituents are also difficult to isolate and identify. Constructing signaling pathways, in particular, pose a challenge as there are various mechanisms by which they can be activated and these mechanisms often overlap. Redundancy, coupled with environmental variables, demonstrates the elaborate ways HSCs develop and maintain their status in the HSC niche. Ambitious efforts are continuously being made to overcome these hurdles, for proper understanding of how various pathways affect stem cell fate can have positive implications for gene therapy.

CONFLICT OF INTEREST

The author confirms that author has no conflict of interest to declare for this publication.

ACKNOWLEDGEMENTS

Declared none.

REFERENCES

[1]   Spangrude GJ, Heimfeld S, Weissman IL. Purification and characterization of mouse hematopoietic stem cells. Science 1988; 241(4861): 58-62.
[http://dx.doi.org/10.1126/science.2898810] [PMID: 2898810]

[2]   Okada S, Nakauchi H, Nagayoshi K, Nishikawa S, Miura Y, Suda T. *In vivo* and *in vitro* stem cell function of c-kit- and Sca-1-positive murine hematopoietic cells. Blood 1992; 80(12): 3044-50.
[PMID: 1281687]

[3]   Matsuzaki Y, Kinjo K, Mulligan RC, Okano H. Unexpectedly efficient homing capacity of purified murine hematopoietic stem cells. Immunity 2004; 20(1): 87-93.
[http://dx.doi.org/10.1016/S1074-7613(03)00354-6] [PMID: 14738767]

[4]   Osawa M, Hanada K, Hamada H, Nakauchi H. Long-term lymphohematopoietic reconstitution by a single CD34-low/negative hematopoietic stem cell. Science 1996; 273(5272): 242-5.
[http://dx.doi.org/10.1126/science.273.5272.242] [PMID: 8662508]

[5]   Yang L, Bryder D, Adolfsson J, *et al.* Identification of Lin(-)Sca1(+)kit(+)CD34(+)Flt3- short-term hematopoietic stem cells capable of rapidly reconstituting and rescuing myeloablated transplant recipients. Blood 2005; 105(7): 2717-23.
[http://dx.doi.org/10.1182/blood-2004-06-2159] [PMID: 15572596]

[6]   Adolfsson J, Borge OJ, Bryder D, *et al.* Upregulation of Flt3 expression within the bone marrow Lin(-)Sca1(+)c-kit(+) stem cell compartment is accompanied by loss of self-renewal capacity. Immunity 2001; 15(4): 659-69.
[http://dx.doi.org/10.1016/S1074-7613(01)00220-5] [PMID: 11672547]

[7]   Morrison SJ, Weissman IL. The long-term repopulating subset of hematopoietic stem cells is deterministic and isolatable by phenotype. Immunity 1994; 1(8): 661-73.
[http://dx.doi.org/10.1016/1074-7613(94)90037-X] [PMID: 7541305]

[8]   Till JE, McCulloch EA, Siminovitch L. A stochastic model of stem cell proliferation, based on the growth of spleen colony-forming cells. Proc Natl Acad Sci USA 1964; 51: 29-36.
[http://dx.doi.org/10.1073/pnas.51.1.29] [PMID: 14104600]

[9]   Metcalf D. Hematopoietic regulators: redundancy or subtlety? Blood 1993; 82(12): 3515-23.
[PMID: 8260692]

[10]   Enver T, Heyworth CM, Dexter TM. Do stem cells play dice? Blood 1998; 92: 348-51. discussion 352.

[11]   Miller CL, Eaves CJ. Expansion *in vitro* of adult murine hematopoietic stem cells with transplantable lympho-myeloid reconstituting ability. Proc Natl Acad Sci USA 1997; 94(25): 13648-53.
[http://dx.doi.org/10.1073/pnas.94.25.13648] [PMID: 9391080]

[12]   Christensen JL, Weissman IL. Flk-2 is a marker in hematopoietic stem cell differentiation: a simple method to isolate long-term stem cells. Proc Natl Acad Sci USA 2001; 98(25): 14541-6.
[http://dx.doi.org/10.1073/pnas.261562798] [PMID: 11724967]

[13]   Audet J, Miller CL, Rose-John S, Piret JM, Eaves CJ. Distinct role of gp130 activation in promoting self-renewal divisions by mitogenically stimulated murine hematopoietic stem cells. Proc Natl Acad Sci USA 2001; 98(4): 1757-62.
[http://dx.doi.org/10.1073/pnas.98.4.1757] [PMID: 11172024]

[14]   Nandurkar HH, Robb L, Tarlinton D, Barnett L, Köntgen F, Begley CG. Adult mice with targeted mutation of the interleukin-11 receptor (IL11Ra) display normal hematopoiesis. Blood 1997; 90(6): 2148-59.
[PMID: 9310465]

[15]   Kimura S, Roberts AW, Metcalf D, Alexander WS. Hematopoietic stem cell deficiencies in mice lacking c-Mpl, the receptor for thrombopoietin. Proc Natl Acad Sci USA 1998; 95(3): 1195-200.
[http://dx.doi.org/10.1073/pnas.95.3.1195] [PMID: 9448308]

[16]   Solar GP, Kerr WG, Zeigler FC, *et al.* Role of c-mpl in early hematopoiesis. Blood 1998; 92(1): 4-10.
[PMID: 9639492]

[17]   Ikuta K, Weissman IL. Evidence that hematopoietic stem cells express mouse c-kit but do not depend on steel factor for their generation. Proc Natl Acad Sci USA 1992; 89(4): 1502-6.
[http://dx.doi.org/10.1073/pnas.89.4.1502] [PMID: 1371359]

[18]   Buza-Vidas N, Antonchuk J, Qian H, *et al.* Cytokines regulate postnatal hematopoietic stem cell expansion: opposing roles of thrombopoietin and LNK. Genes Dev 2006; 20(15): 2018-23.
[http://dx.doi.org/10.1101/gad.385606] [PMID: 16882979]

[19]   Borge OJ, Ramsfjell V, Veiby OP, Murphy MJ Jr, Lok S, Jacobsen SE. Thrombopoietin, but not erythropoietin promotes viability and inhibits apoptosis of multipotent murine hematopoietic progenitor cells *in vitro.* Blood 1996; 88(8): 2859-70.
[PMID: 8874182]

[20]   Pestina TI, Cleveland JL, Yang C, Zambetti GP, Jackson CW. Mpl ligand prevents lethal myelosuppression by inhibiting p53-dependent apoptosis. Blood 2001; 98(7): 2084-90.
[http://dx.doi.org/10.1182/blood.V98.7.2084] [PMID: 11567994]

[21]   Seita J, Ema H, Ooehara J, *et al.* Lnk negatively regulates self-renewal of hematopoietic stem cells by modifying thrombopoietin-mediated signal transduction. Proc Natl Acad Sci USA 2007; 104(7): 2349-54.
[http://dx.doi.org/10.1073/pnas.0606238104] [PMID: 17284614]

[22]   Takaki S, Sauer K, Iritani BM, *et al.* Control of B cell production by the adaptor protein lnk. Definition Of a conserved family of signal-modulating proteins. Immunity 2000; 13(5): 599-609.
[http://dx.doi.org/10.1016/S1074-7613(00)00060-1] [PMID: 11114373]

[23]   Tong W, Zhang J, Lodish HF. Lnk inhibits erythropoiesis and Epo-dependent JAK2 activation and

downstream signaling pathways. Blood 2005; 105(12): 4604-12.
[http://dx.doi.org/10.1182/blood-2004-10-4093] [PMID: 15705783]

[24]   Velazquez L, Cheng AM, Fleming HE, *et al.* Cytokine signaling and hematopoietic homeostasis are disrupted in Lnk-deficient mice. J Exp Med 2002; 195(12): 1599-611.
[http://dx.doi.org/10.1084/jem.20011883] [PMID: 12070287]

[25]   Ema H, Sudo K, Seita J, *et al.* Quantification of self-renewal capacity in single hematopoietic stem cells from normal and Lnk-deficient mice. Dev Cell 2005; 8(6): 907-14.
[http://dx.doi.org/10.1016/j.devcel.2005.03.019] [PMID: 15935779]

[26]   Radtke F, Wilson A, Mancini SJ, MacDonald HR. Notch regulation of lymphocyte development and function. Nat Immunol 2004; 5(3): 247-53.
[http://dx.doi.org/10.1038/ni1045] [PMID: 14985712]

[27]   Milner LA, Kopan R, Martin DI, Bernstein ID. A human homologue of the Drosophila developmental gene, Notch, is expressed in CD34+ hematopoietic precursors. Blood 1994; 83(8): 2057-62.
[PMID: 7512837]

[28]   Karanu FN, Murdoch B, Gallacher L, *et al.* The notch ligand jagged-1 represents a novel growth factor of human hematopoietic stem cells. J Exp Med 2000; 192(9): 1365-72.
[http://dx.doi.org/10.1084/jem.192.9.1365] [PMID: 11067884]

[29]   Karanu FN, Murdoch B, Miyabayashi T, *et al.* Human homologues of Delta-1 and Delta-4 function as mitogenic regulators of primitive human hematopoietic cells. Blood 2001; 97(7): 1960-7.
[http://dx.doi.org/10.1182/blood.V97.7.1960] [PMID: 11264159]

[30]   Varnum-Finney B, Brashem-Stein C, Bernstein ID. Combined effects of Notch signaling and cytokines induce a multiple log increase in precursors with lymphoid and myeloid reconstituting ability. Blood 2003; 101(5): 1784-9.
[http://dx.doi.org/10.1182/blood-2002-06-1862] [PMID: 12411302]

[31]   Vas V, Szilágyi L, Pálóczi K, Uher F. Soluble Jagged-1 is able to inhibit the function of its multivalent form to induce hematopoietic stem cell self-renewal in a surrogate *in vitro* assay. J Leukoc Biol 2004; 75(4): 714-20.
[http://dx.doi.org/10.1189/jlb.1003462] [PMID: 14742638]

[32]   Ohishi K, Varnum-Finney B, Bernstein ID. Delta-1 enhances marrow and thymus repopulating ability of human CD34(+)CD38(-) cord blood cells. J Clin Invest 2002; 110(8): 1165-74.
[http://dx.doi.org/10.1172/JCI0216167] [PMID: 12393852]

[33]   Delaney C, Varnum-Finney B, Aoyama K, Brashem-Stein C, Bernstein ID. Dose-dependent effects of the Notch ligand Delta1 on *ex vivo* differentiation and *in vivo* marrow repopulating ability of cord blood cells. Blood 2005; 106(8): 2693-9.
[http://dx.doi.org/10.1182/blood-2005-03-1131] [PMID: 15976178]

[34]   Suzuki A, Raya A, Kawakami Y, *et al.* Nanog binds to Smad1 and blocks bone morphogenetic protein-induced differentiation of embryonic stem cells. Proc Natl Acad Sci USA 2006; 103(27): 10294-9.
[http://dx.doi.org/10.1073/pnas.0506945103] [PMID: 16801560]

[35]   Kunisato A, Chiba S, Nakagami-Yamaguchi E, *et al.* HES-1 preserves purified hematopoietic stem

cells *ex vivo* and accumulates side population cells *in vivo.* Blood 2003; 101(5): 1777-83.
[http://dx.doi.org/10.1182/blood-2002-07-2051] [PMID: 12406868]

[36]   Stier S, Cheng T, Dombkowski D, Carlesso N, Scadden DT. Notch1 activation increases hematopoietic stem cell self-renewal *in vivo* and favors lymphoid over myeloid lineage outcome. Blood 2002; 99(7): 2369-78.
[http://dx.doi.org/10.1182/blood.V99.7.2369] [PMID: 11895769]

[37]   Varnum-Finney B, Xu L, Brashem-Stein C, *et al.* Pluripotent, cytokine-dependent, hematopoietic stem cells are immortalized by constitutive Notch1 signaling. Nat Med 2000; 6(11): 1278-81.
[http://dx.doi.org/10.1038/81390] [PMID: 11062542]

[38]   Calvi LM, Adams GB, Weibrecht KW, *et al.* Osteoblastic cells regulate the haematopoietic stem cell niche. Nature 2003; 425(6960): 841-6.
[http://dx.doi.org/10.1038/nature02040] [PMID: 14574413]

[39]   Mancini SJ, Mantei N, Dumortier A, Suter U, MacDonald HR, Radtke F. Jagged1-dependent Notch signaling is dispensable for hematopoietic stem cell self-renewal and differentiation. Blood 2005; 105(6): 2340-2.
[http://dx.doi.org/10.1182/blood-2004-08-3207] [PMID: 15550486]

[40]   Willert K, Brown JD, Danenberg E, *et al.* Wnt proteins are lipid-modified and can act as stem cell growth factors. Nature 2003; 423(6938): 448-52.
[http://dx.doi.org/10.1038/nature01611] [PMID: 12717451]

[41]   Reya T, Duncan AW, Ailles L, *et al.* A role for Wnt signalling in self-renewal of haematopoietic stem cells. Nature 2003; 423(6938): 409-14.
[http://dx.doi.org/10.1038/nature01593] [PMID: 12717450]

[42]   Kirstetter P, Anderson K, Porse BT, Jacobsen SE, Nerlov C. Activation of the canonical Wnt pathway leads to loss of hematopoietic stem cell repopulation and multilineage differentiation block. Nat Immunol 2006; 7(10): 1048-56.
[http://dx.doi.org/10.1038/ni1381] [PMID: 16951689]

[43]   Cobas M, Wilson A, Ernst B, *et al.* Beta-catenin is dispensable for hematopoiesis and lymphopoiesis. J Exp Med 2004; 199(2): 221-9.
[http://dx.doi.org/10.1084/jem.20031615] [PMID: 14718516]

[44]   Duncan AW, Rattis FM, DiMascio LN, *et al.* Integration of Notch and Wnt signaling in hematopoietic stem cell maintenance. Nat Immunol 2005; 6(3): 314-22.
[http://dx.doi.org/10.1038/ni1164] [PMID: 15665828]

[45]   Trowbridge JJ, Xenocostas A, Moon RT, Bhatia M. Glycogen synthase kinase-3 is an *in vivo* regulator of hematopoietic stem cell repopulation. Nat Med 2006; 12(1): 89-98.
[http://dx.doi.org/10.1038/nm1339] [PMID: 16341242]

[46]   Sitnicka E, Ruscetti FW, Priestley GV, Wolf NS, Bartelmez SH. Transforming growth factor beta 1 directly and reversibly inhibits the initial cell divisions of long-term repopulating hematopoietic stem cells. Blood 1996; 88(1): 82-8.
[PMID: 8704205]

[47]   Batard P, Monier MN, Fortunel N, *et al.* TGF-(beta)1 maintains hematopoietic immaturity by a

reversible negative control of cell cycle and induces CD34 antigen up-modulation. J Cell Sci 2000; 113(Pt 3): 383-90.
[PMID: 10639326]

[48]   Garbe A, Spyridonidis A, Möbest D, Schmoor C, Mertelsmann R, Henschler R. Transforming growth factor-beta 1 delays formation of granulocyte-macrophage colony-forming cells, but spares more primitive progenitors during *ex vivo* expansion of CD34+ haemopoietic progenitor cells. Br J Haematol 1997; 99(4): 951-8.
[http://dx.doi.org/10.1046/j.1365-2141.1997.4893291.x] [PMID: 9432049]

[49]   Larsson J, Blank U, Helgadottir H, *et al.* TGF-beta signaling-deficient hematopoietic stem cells have normal self-renewal and regenerative ability *in vivo* despite increased proliferative capacity *in vitro.* Blood 2003; 102(9): 3129-35.
[http://dx.doi.org/10.1182/blood-2003-04-1300] [PMID: 12842983]

[50]   Larsson J, Blank U, Klintman J, Magnusson M, Karlsson S. Quiescence of hematopoietic stem cells and maintenance of the stem cell pool is not dependent on TGF-beta signaling *in vivo.* Exp Hematol 2005; 33(5): 592-6.
[http://dx.doi.org/10.1016/j.exphem.2005.02.003] [PMID: 15850837]

[51]   Nishita M, Hashimoto MK, Ogata S, *et al.* Interaction between Wnt and TGF-beta signalling pathways during formation of Spemann's organizer. Nature 2000; 403(6771): 781-5.
[http://dx.doi.org/10.1038/35001602] [PMID: 10693808]

[52]   Labbé E, Letamendia A, Attisano L. Association of Smads with lymphoid enhancer binding factor 1/T cell-specific factor mediates cooperative signaling by the transforming growth factor-beta and wnt pathways. Proc Natl Acad Sci USA 2000; 97(15): 8358-63.
[http://dx.doi.org/10.1073/pnas.150152697] [PMID: 10890911]

[53]   Itoh F, Itoh S, Goumans MJ, *et al.* Synergy and antagonism between Notch and BMP receptor signaling pathways in endothelial cells. EMBO J 2004; 23(3): 541-51.
[http://dx.doi.org/10.1038/sj.emboj.7600065] [PMID: 14739937]

[54]   Schofield R. The relationship between the spleen colony-forming cell and the haemopoietic stem cell. Blood Cells 1978; 4(1-2): 7-25.
[PMID: 747780]

[55]   Wilson A, Trumpp A. Bone-marrow haematopoietic-stem-cell niches. Nat Rev Immunol 2006; 6(2): 93-106.
[http://dx.doi.org/10.1038/nri1779] [PMID: 16491134]

[56]   Stier S, Ko Y, Forkert R, *et al.* Osteopontin is a hematopoietic stem cell niche component that negatively regulates stem cell pool size. J Exp Med 2005; 201(11): 1781-91.
[http://dx.doi.org/10.1084/jem.20041992] [PMID: 15928197]

[57]   Nilsson SK, Johnston HM, Whitty GA, *et al.* Osteopontin, a key component of the hematopoietic stem cell niche and regulator of primitive hematopoietic progenitor cells. Blood 2005; 106(4): 1232-9.
[http://dx.doi.org/10.1182/blood-2004-11-4422] [PMID: 15845900]

[58]   Adams GB, Chabner KT, Alley IR, *et al.* Stem cell engraftment at the endosteal niche is specified by the calcium-sensing receptor. Nature 2006; 439(7076): 599-603.
[http://dx.doi.org/10.1038/nature04247] [PMID: 16382241]

[59] Parmar K, Mauch P, Vergilio JA, Sackstein R, Down JD. Distribution of hematopoietic stem cells in the bone marrow according to regional hypoxia. Proc Natl Acad Sci USA 2007; 104(13): 5431-6. [http://dx.doi.org/10.1073/pnas.0701152104] [PMID: 17374716]

[60] Ito K, Hirao A, Arai F, *et al.* Regulation of oxidative stress by ATM is required for self-renewal of haematopoietic stem cells. Nature 2004; 431(7011): 997-1002. [http://dx.doi.org/10.1038/nature02989] [PMID: 15496926]

[61] Ito K, Hirao A, Arai F, *et al.* Reactive oxygen species act through p38 MAPK to limit the lifespan of hematopoietic stem cells. Nat Med 2006; 12(4): 446-51. [http://dx.doi.org/10.1038/nm1388] [PMID: 16565722]

[62] Bowie MB, McKnight KD, Kent DG, McCaffrey L, Hoodless PA, Eaves CJ. Hematopoietic stem cells proliferate until after birth and show a reversible phase-specific engraftment defect. J Clin Invest 2006; 116(10): 2808-16. [http://dx.doi.org/10.1172/JCI28310] [PMID: 17016561]

[63] Gluckman E, Broxmeyer HA, Auerbach AD, *et al.* Hematopoietic reconstitution in a patient with Fanconi's anemia by means of umbilical-cord blood from an HLA-identical sibling. N Engl J Med 1989; 321(17): 1174-8. [http://dx.doi.org/10.1056/NEJM198910263211707] [PMID: 2571931]

[64] Attar EC, Scadden DT. Regulation of hematopoietic stem cell growth. Leukemia 2004; 18(11): 1760-8. [http://dx.doi.org/10.1038/sj.leu.2403515] [PMID: 15457182]

[65] Antonchuk J, Sauvageau G, Humphries RK. HOXB4-induced expansion of adult hematopoietic stem cells *ex vivo.* Cell 2002; 109(1): 39-45. [http://dx.doi.org/10.1016/S0092-8674(02)00697-9] [PMID: 11955445]

[66] Amsellem S, Pflumio F, Bardinet D, *et al. Ex vivo* expansion of human hematopoietic stem cells by direct delivery of the HOXB4 homeoprotein. Nat Med 2003; 9(11): 1423-7. [http://dx.doi.org/10.1038/nm953] [PMID: 14578882]

[67] Krosl J, Austin P, Beslu N, Kroon E, Humphries RK, Sauvageau G. In *vitro* expansion of hematopoietic stem cells by recombinant TAT-HOXB4 protein. Nat Med 2003; 9(11): 1428-32. [http://dx.doi.org/10.1038/nm951] [PMID: 14578881]

Developmental and Stem Cell Biology in Health and Disease, 2016, 49-81

CHAPTER 4

Stem Cell Biology: Flies As Models and Examples

Nahla M. Shoukry[1,*], **Elham M. Youssef Elabd**[2], **Hani S. Hafez**[1] and **Noha M. Osman**[2,3,4]

[1] *Zoology Department, Faculty of Science, Suez University, Suez, Egypt*

[2] *National Research Center, Egypt*

[3] *Department of Molecular and Computational biology, University of Southern California, USA*

[4] *Washington State University, Pullman, Washington, USA*

Abstract: Drosophila is a genus of flies of the family Drosophilidae. It comprises of 1579 described species. Stem cell present in stem cell niche and interacts to regulate cell fate. Many factors act on embryonic stem cells during embryonic development to alter gene expression, and induce their proliferation or differentiation for the development of the fetus or larvae.

Stem cell niches maintain adult stem cells in a quiescent state. After tissue injury, the surrounding micro-environment induce stem cells to promote either self-renewal or differentiation to new tissues. Various factors are important to coordinate stem cell characteristics within the niche: cell-cell interactions between stem cells, as well as interactions between stem cells and neighbouring differentiated cells, interactions between stem cells and adhesion molecules, extracellular matrix components, the oxygen tension, growth factors, cytokines, and the physicochemical nature of the environment including the pH, ionic strength (*e.g.* Ca^{2+} concentration) and metabolites, like ATP, are also important. During development, stem cells and niche may induce each other and alternately maintain each other during adulthood.

Keywords: Germline stem cell, Niche signaling, Niche morphogenesis, Somatic stem cell, siRNA.

---

* **Correspondence author Nahla M. Shoukry:** Zoology Department, Faculty of Science, Suez University, Suez, Egypt; E-mails: nahla.m.shoukry@gmail.com, Nahla.Shokry@suezuniv.edu.eg

Ahmed El-Hashash (Ed)

BASICS OF STEM CELLS

Stem cells are capable to develop into many different cell types in the body during early life and growth.

Moreover, they act as a sort of internal repair system in many tissues by dividing without limit to replace worn out or damaged tissues as long as the person or animal is still alive. When a stem cell divides, each new cell has the potential either to remain a stem cell or differentiate to another type of cell with a more specialized function, such as a muscle cell, a red blood cell, or a brain cell. Stem cells are characterized from other cell types by two important features. First, they are unspecialized cells capable of self-renewing through symmetric cell division, sometimes after long periods of inactivity. Second, under certain physiologic or experimental conditions, they can be induced to differentiate to tissue- or organ-specific cells with special functions, such as gut and bone marrow. Most recently, scientists primarily worked with two kinds of stem cells from animals and humans: embryonic stem cells and non-embryonic "somatic" or "adult" stem cells. Scientists discovered methods to derive embryonic stem cells from early mouse embryos more than 30 years ago. In 1998, the detailed study of the mouse stem cells' biology led to the discovery of a method to derive stem cells from human embryos and grow them in the laboratory. These cells are called human embryonic stem cells. The embryos used in these studies were created for reproductive purposes through *in vitro* fertilization. When they were no longer needed for that purpose, they were donated for research with the informed consent of the donor. In 2006, researchers made another breakthrough in stem cell research by identifying conditions that would allow some specialized adult cells to be genetically "reprogrammed" to be a stem cell-like state. This new type of stem cell, called induced pluripotent stem cells (iPSCs), will be discussed in a later section of this document. Stem cells are important for living organisms for many reasons. In the 3- to 5-day-old embryo, called a blastocyst, the inner cell mass give rise to the entire body of the organism, including all of the specialized cell types and organs such as the heart, lungs, skin, sperm, eggs and other tissues. In some adult tissues, such as bone marrow, muscle, and brain, adult stem cells repair and regenerate cells that are lost through normal wear and tear, injury, or disease.

NICHE REGULATION IN DROSOPHILA GERMLINE STEM CELLS

Germline stem cells produce progeny that proliferate through several rounds of divisions, and then enter meiosis. Although this transit-amplifying (TA) strategy is widely conserved in spermatogenesis, the germline mitotic proliferation program appears very different from the perspective of the female. Although female germline stem cells produce daughter cells that initiate differentiation with rounds of mitotic division, producing cysts of interconnected mitotic sisters in both Drosophila and mice [1, 2]. These divisions are not TA, because the number of oocytes is not amplified. The TA program likely underlies germ cell behavior in both sexes, but the strategy has been modified during evolution in the female so that most of the mitotic sisters sacrifice their contents to build the specialized oocyte. This sex difference goes along with two other main differences in germ cells between the sexes [3]. The female gamete is larger and usually contains extensive components to carry out the start of embryogenesis [4]. The meiotic divisions take place after the major events of gametogenesis in females but before the major morphogenetic events that lead to the final form of the gametes in males [5]. A classic example occurs in *C. elegans*, where a large somatic cell at the distal tip of the gonad maintains a syncytial population of mitotically proliferating germ cell nuclei through close-range instruction *via* Notch pathway signaling [6]. The subpopulation of the germline mitotic nuclei lying closest to the distal tip likely corresponds to germline stem cells [7], whereas the mitotic nuclei lying further from the distal tip likely correspond to the TA population.

STEM CELL NICHE IN THE OVARIAN GERMLINE OF DROSOPHILA

Drosophila females maintain germ cell production throughout adult life by establishing a niche at the tip of each ovariole that prevents some primordial germ cells (PGCs) from differentiating. The ability to form niches appears to distinguish species with female germline stem cell (GSCs) from those without rather than basic differences in germ cell differentiation. Niches in the ovary arise near the end of larval development and upload PGCs, which are produced in excess [8], just prior to the onset of global differentiation signals. PGCs that make it into the niche are protected and become GSCs, whereas identical PGCs outside begin development [9]. In addition to sequestering a supply of undifferentiated germ cells, niches program subsequent GSC divisions to be asymmetric by providing only enough room for one of the two daughters to remain inside. In simple manner, GSC niches generate a lifelong supply of germ cells for egg production. The niche itself is built from three somatic cell types: terminal filament cells located at the anterior of each ovariole, cap cells, which actually contact GSCs and hold them in the niche, and escort cells, which wrap thin processes around GSCs and prevent direct GSC–GSC contact. Genes promoting PGC/GSC differentiation are modulated by changes at the level of chromatin structure, transcription, and RNA processing, but especially at the level of translation. The conserved NANOS and PUMILIO translational repressors are essential to prevent premature germ cell differentiation [10 - 12], indicating that messenger RNAs (mRNAs) capable of promoting cyst formation and oogenesis are present within PGCs. These mRNAs are controlled by proteins, including VASA, which localize them within large silencing ribo-nucleoprotein particles (RNPs), promote CCR4-meditated de-adenylation, and which influence their binding to 40S small ribosomal subunits in conjunction with translation initiation factors such as eIF4E [12 - 15]. Indeed, the perinuclear nuage, stage- specific nuclear and cytoplasmic germ granules, and chromatoid bodies that type germ cells throughout the metazoa, are manifestations of abundant RNA processing complexes participating in germ cell regulation [16 - 20].

Drosophila ovarian niches control the onset of germ cell development by spatially modulating expression of a high-level differentiation inducer encoded by bag-of-marbles (bam). Cap cells and possibly anterior escort cells activate JAK-STAT signaling upon receipt of the unpaired ligand from terminal filament cells and respond by producing the BMP ligands Dpp and Gbb [21, 22]. Consequently, germ cells located within a small zone adjacent to the cap cells are positioned to receive a strong BMP signal. Activation of BMP receptors in GSCs directly represses bam transcription [23]. Only upon leaving the niche do they escape BMP-mediated repression, up- regulate bam expression, and become cystoblasts, *i.e.,* cells that will found new 16-cell cysts. Moreover, bam can initiate development equally well within larval ovarian PGCs, but premature differentiation is prevented by gonad-wide BMP signaling during these stages [9, 24]. Interestingly, the bam gene is not widely conserved and shows signs of recent positive selection [25], raising the possibility that it arose in concert with the evolution of Drosophila female GSCs. A key niche requirement is to limit high level Dpp/Gbb signaling to a one-cell-wide zone. The DALLY heparin sulfate glycoprotein, which accumulates at high levels only around the cap cells, binds to and stabilizes Dpp and may be necessary to effectively activate BMP signaling [26, 27]. EGFR signaling from early germ cells to escort cells mediated by the membrane protease STET influences DALLY levels or function [28]. The ability of germ cells downstream from the niche to receive BMP signals is further controlled by negative feedback from the BMP-induced E3 ubiquitin ligase SMURF and the Ser/Thr kinase FUSED, which promotes turnover of the BMP receptor THICK VEINS (Tkv) [29, 30]. All these mechanisms ensure that once outside the niche, bam transcription will be promptly and stably activated.

STEM CELLS WITHOUT A NICHE: DROSOPHILANEUROBLASTS

During brain development, neural stem cells proliferate in a spatially and temporally regulated fashion producing enormous and diverse neurons which will drive the complex behavior of adult animals. So, the limited number of neural stem cells generating different types of brain cell is still obscured issue. On one side, neurons arise from asymmetric divisions of a progenitor daughter cell that retains self-renewal capacity and the other is committed to neural differentiation. Different types of neurons are generated over time and this is facilitated by stereotyped transcriptional changes in the progenitor cell that follow a precise temporal order. Cell cycle entry and exit of neural stem cells are coordinated with developmental time to ensure that the right neurons are created at the right time and to prevent the formation of brain tumors. Furthermore, unlike other organs, the brain is spared from growth restrictions under starvation probably because a full complement of neurons is needed for the brain to function. Surprisingly, all of these key features can be recapitulated in neuroblasts, the stem cells found in the developing brain of the simple invertebrate Drosophila melanogaster.

The simplicity of Drosophila development and the sophisticated genetic tools available for studying Drosophila have allowed obtaining insights into stem cell biology that would not be possible in a vertebrate model system. We then review how the study of stem cell proliferation and regulation has proven useful for understanding how stem cell tumors are originated, highlighting how tumor neuroblasts differ from wild-type neuroblasts and how they might bypass temporal and nutrient- sensing regulatory mechanisms. Finally, we discuss how the mechanisms that regulate neural stem cells in flies parallel those observed in vertebrates.

DEVELOPMENT OF NEUROBLASTS IN DROSOPHILA

Neuroblasts (NBs) are first formed during the embryonic stages (stages 9 to 10) of Drosophila development. NBs delaminate from a neuro-epithelium located in the ventro-lateral region of the embryo and start dividing shortly afterwards to generate neurons and glia. Embryonic NBs are specified in a process called lateral inhibition in which Notch/Delta signaling refines the expression of proneural genes to individual cells [5, 31]. NBs undergo multiple asymmetric divisions [32], giving rise to another NB and a smaller ganglion mother cell (GMC), which divides once to produce neurons and/ or glial cells. Embryonic NB divisions produce all neurons that will form the larval central nervous system (CNS) but only 10% of the cells in the adult CNS [33, 34]. Most NBs in the abdominal regions of the embryo are eliminated through programmed cell death after completing their neuronal lineages [35]. In the cephalic and thoracic regions, however, NBs arrest their cell cycle and exit from G1 into a G0- like quiescent state. Around 8-10 hours after larval hatching, during the late 1st instar stage, the NBs start exiting quiescence and reenter mitosis. This second wave of neurogenesis is responsible for the formation of 90% of the neurons in the adult CNS. Neurogenesis continues throughout larval stages into pupal stages, at which point the NBs exit from the cell cycle and disappear [36 - 38]. The larval brain in particular has been used extensively to study how NB lineage progression is regulated. Unlike embryonic NBs, larval NBs re-grow to their original size after each cell division and are capable of dividing hundreds of times. Based on their position in the brain and their lineage characteristics, we can differentiate abdominal and thoracic neuroblasts in the ventral nerve cord (VNC) and optic lobe NBs in the brain lobes. Ventral nerve cord and central brain NBs originate from embryonic NBs, optic lobe NBs arise only during larval stages [39, 40].

PROGENITOR CELL NICHE IN DROSOPHILA HEMATOPOIETIC PRECURSOR CELLS

The blood cells mammalian hematopoiesis gives rise to long-term reconstituting hematopoietic stem cell (HSCs) that generate short-term repopulating HSCs [41] with some types of progenitors. Hematopoietic stem cell gives rise to all differentiated blood cells in adult circulation, such as lymphoid, myeloid, and erythroid cells [3]. Development of the HSC requires Runx-1 [42] producing differentiating myeloid and lymphoid cells [43, 44]. Hematopoietic lineages in Drosophila are not as diverse as in vertebrates and all present hemocytes are of ancient myeloid lineage. While there are no erythroid and lymphoid cell types, hemocytes with myeloid macrophage properties are represented [45, 46]. Hemocytes remain in circulation during larval development and in adults due to its role in immunity, wound repair, and apoptosis/phagocytosis [46 - 49]. The blood cell types in Drosophila can be classified into three major classes. Plasmatocytes are phagocytic macrophage like cells that comprise >90% of the circulating population of hemocytes in wild-type larvae. Crystal cells possess paracrystalline inclusions of prophenoloxidase similar in its melanization function to the vertebrate Tyrosinase, while lamellocytes engulf larger particles of infection. Lamellocytes and crystal cellsform about less than 5%. Hematopoietic system contains a number of conserved signaling pathways and transcription factors between the vertebrate and the Drosophila hematopoietic systems [45, 50, 51].Hematopoiesis in most vertebrate embryos occurs in sequential primitive and definitive waves [52]. In mammals primitive hematopoiesis gives rise to transient populations of progenitors that differentiate into erythrocytes and macrophages [52, 53] followed by definitive hematopoiesis which gives rise to HSCs generating full range of blood cell types in embryo and throughout adulthood [52].

In Drosophila as well, hematopoiesis occurs in two phases: one originating from the anterior head mesoderm giving rise to an early set of blood cells in the embryo, while additional definitive blood cell production takes place in the larval lymph gland, which contributes to the adult [54].

SITES OF TRANSIENT EMBRYONIC HEMATOPOIESIS

Drosophila intra-embryonic sites of early blood formation, particularly from head mesodermand, theinitial wave of hemocyte differentiation also occur in the head mesoderm region during the early embryonic period [55]. Glial cells missing Gcm is required for the specification of plasmatocytes [56, 57],and the Runx-like transcription factor Lozenge (Lz) is necessary for crystal cell development [57].

In mammals, microglia serve as phagocytic functions in the brain similar to those of macrophages in the periphery [58]. The primitive macrophages in Drosophila arising from the head mesoderm are akin to the monocyte-like mammalian microglial population, which originates independently of the bone marrow-derived monocyte precursor [58]. The first hematopoietic organ in both mouse and human is the yolk sac, producing a well- established role in the generation of transient hematopoietic populations for the immediate needs of the embryo, including primitive red cells required for oxygen transport [59], macrophages for tissue remodeling and defense [60], and unique primitive megakaryocytes [61]. After this so- called primitive hematopoiesis, a burst of production of multipotential myeloerythroid progenitors occurs [62]. They are the first hematopoietic cells that seed the liver, where they give rise to definitive red cells and myeloid cells. The literature on the exact nature of these cells and their role has been controversial [62].

These progenitors are classified as definitive, as their erythroid progeny express adult type globins. However, they should perhaps be named transient or short-term definitive progenitors since their ability to contribute to adult hematopoiesis has been doubtful. Both direct transplantation into adult recipients and explant cultures of early yolk sac tissues have failed to verify de novo HSC generation in the yolk sac. In contrast, other studies have suggested that yolk sac cells may contribute to adult hematopoiesis if injected into fetal environments. More recently, an estrogen- inducible mouse model was utilized to permanently label Runx1-expressing cells prior to circulation, and contribution of the labeled cells to adult hematopoiesis led to the conclusion that HSC generation starts in the yolk sac [63]. It is therefore possible that the negative results in adult transplantation assays reflect immaturity of the nascent yolk sac hematopoietic cells, which may be yet unable to engraft and survive in adult niches. Alternatively, these cells may represent a transient precursor population that shares only some characteristics with adult-type definitive hematopoietic cells, but do not contribute to adult hematopoiesis. These scenarios are not mutually exclusive; it is plausible that the yolk sac generates three waves of hematopoietic cells: primitive hematopoietic cells, transient definitive progenitors, and definitive HSCs. Although it is yet unclear whether all the waves of hematopoiesis in the yolk sac have a common ancestry, it has been shown that mesodermal cells that migrate through the primitive streak have both primitive and definitive hematopoietic as well as endothelial potential, suggesting that they represent hemangioblasts that give rise to yolk sac hematopoietic cells [64].

FUNCTIONS OF EXTRACELLULAR MATRIX AND CELL ADHESION PROTEINS IN NICHE MORPHOGENESIS CHARACTERISTICS

Stem cell niches consist of supporting cells and associated extracellular matrix (ECM). The spatial organization of stem cells into a niche is a key factor for growth and continual tissue renewal during development, and regeneration. How possible the system of stem cells and associated neighboring cells become organized? Properly regulated cell adhesion is believed to link several major developmental signaling path ways which are crucial for cell morphogenesis. As a matter of fact, the equilibrium of cell-cell adhesion property can orchestrate both changes in arrangements and cellular behavior including mobilization, developing of cellular compositions like epithelial sheets, sorting and isolating of cell population [65]. Work largely on Drosophila has started to reveal the molecular mechanisms that participate in the maintenance of stem cell identity .Two fundamental mechanisms, cell signaling and adhesive mechanisms, govern the process. The elements in adhesion mechanism, DE-cadherin and Armadillo/β-catenin, are concentrated in the junction between GSCs and cap as junctions. The existence of these kind of junctions is commonly shown by concentration of their components at high levels in the surfaces of the cells involved [65, 66]. Moreover, other adhesion molecules which are accumulated between GSCs and stromal cells [67 - 70], indicate their essential role in the stem cell organization. In fact, mutant fGSC clones that were lost very quickly in the disturbance of adherens junctions confirm the fact that proper attachment between both type of cells, germline stem cells and niche cells is critical for GSCs structure, function and self-renewal mechanisms [69, 71, 72]. Moreover, recent findings suggest that niche-germline cell adhesion that are mediated by Cadherin probably have GSCs [73, 74] and Rab11-mediated trafficking of DE-Cadherin should have a role in a competitive relationship between normal Drosophila ovarian germline and those with differentiation defects for niche occupancy. This occurs by up regulating E-cadherin expression [75]. However, a lack of expression of DE-Cadherin has been specifically related to the loss of the maintenance of stem cell and the integrity of their niche in male and female gonads during aging process [76, 77]. Adherens junctions also seem essential to maintain further cells throughout the ovary such as the somatic follicle stem cells [78]. Recently, various studies have focused on the mechanism of stem cells maintenance in their niches. They indicate that niche of male GSC probably forms during the process of embryogenesis , as a group of cells from the anterior somatic gonadal precursors (SGPs)starts to accumulate Fasciclin III, DE- and N- Cadherin adhesion proteins, similar to those of adult hub cells [79]. During the final stages of embryogenesis a small number of primordial germ cells (PGCs) PGCs are organized into a characteristic rosette-like structure [79]. Moreover, it is believed that GTPase signaling plays an important role in adhesion of niche cells. Mutation of Rap-GEF seems to disturb the induction of adherent elements among hub cells and mGSCs. Additionally germline stem cells were observed during larva development in Rap-GEF mutants. Rap GTPase may act primarily on the hub cells and its lost is followed by a mechanism of compensation through over expressing of DE-Cadherin in mGSCs. This indicates that cell attachment is essential for niche function [72]. The Extracellular Matrix (ECM) is thought to be an important component of niche because in many cases, stem cells such as ISCs and FSCs, directly contact the ECM. They lack direct association with identified stromal cells and are located adjacent to the basement membrane. Thus the identification of such putative stem cell niches [80, 81] has clarified the essential role of stem cell attachment to the basement membrane and the indirect cell-to-cell communication that is mediated *via* the Extracellular Matrix (ECM). During the early stages of embryo development, ECM elements gather surrounding gonads of males and by the latter stage of embryogenesis the hub cells appear to communicate with the ECM by adopting their anterior position [82]. It has also been shown that integrin is critical for appropriate hub cells attachment to the ECM. Moreover defect in integrin lead to an improper position of embryonic hub cells during gonad morphogenesis. Recently, it has been suggested that ECM has a key role not only in niche morphology but in regulation of niche signaling as well as the IV collagen extracellular matrix proteins are essential in the formation of BMP gradient in the niche of fGSC [21]. However, the mechanism of ECM establishment and maintenance in niches with unstable stromal cell associations still remains to be elucidated.

DOES IT DIFFER BETWEEN MALE AND FEMALE FLIES? DROSOPHILA

Germline stem cells in male and female drosophila have become an important model system in studying the communication between stem cells and their niche. Although the GSC differentiation pathways that both testis and ovary share are common, their communication in the corresponding organs is different. Gene analysis studies have determined individual signaling pathways to define permissive zone for stem cell self-renewal [83, 84]. In GSC niche in Drosophila ovaries the asymmetric cell division is likely to correlate another contributing factors regenerating from the niche. BMPs and Piwi, are produced from cap cells to regulate GSC division in ovary through intracellular signaling pathways. Dpp and Gbb function in activation of BMP together with the Otfin protein [85]. Further, JAK-STAT signaling is crucial for self- renewal of drosophila testis. Additionally, Unpaired (upd) contributes in STAT activation and expression of Zfh1 [86]. The greatly expanded studies on Drosophila stem cells and niches indicate that there are strong similarities in signaling pathways involved in both mGSC and fGSC niches regardless differences in anatomy [87].

Recently, it has been described that the JAK/STAT signaling pathway plays an important role in niche function in ovary [21, 88, 89]. Conversely, when JAK/STAT signaling is increased by driving the unpaired ligand upd in germarial somatic cells, the number of cysts increased, and the germarium became disorganized, demonstrating that this pathway is enough keep the female germline stem cell niche retained in the ovary. On the other hand, such signaling path way is not required for the maintenance of male stem cell in testis [21, 88], which means that it may mediate indirect effects. Upd2 promote mGSC tumor formation significantly and requires increasing in the expression of dpp and the activation of BMP signaling pathway [21]. Furthermore, JAK/STAT signaling in ESCs is probably required for their maintenance [88]. In agreement with this, the loss of hop is enough to induce GSC differentiation [21, 89]. The communication between support cells and stem cells *via* JAK/STAT and BMP signaling pathways in the fGSC niche has shed light on whether male and female GSC niches develop and operate using similar regulatory mechanisms. In fact, JAK/STAT signaling pathway is essential in maintaining mGSCs self-renewal as well. Furthermore, this signaling does not act alone, but in conjunction with BMP signaling to ensure the GSC renewal. A strong suppression of bam expression follows abnormal activation of the BMP signaling pathway in the germ line of Drosophila. However, this is not enough to induce the formation of mGSC tumor [90, 91]. It also appears that tumors induced by upd are unable to express bam [37]. Since signals controlled by JAK\STAT and BMP in the germ line repress bam expression, it is possible that these pathways interact in germ line cells. The association of BMP pathway activation with upd- induced tumors is yet to be reported. BMP seems to have indirect effect on gonads since their gradient ligands are not visualized. The regulatory function of BMP and Bam in turning from amplifying the number of germ cells in the cyst to initiating spermatogenesis [91, 92] indicates that BMP ligands probably distributed more in the testis than in the ovary. This probably explains the reason why JAK\STAT signaling is a limiting key defining the male GSC niche.

Several signaling pathways have been identified in regulation of stem cell self-renewal. However, it is clear that different molecules probably function by underlying mechanisms in various types of stem cell niches .Among these underlying mechanisms: RNA expression control and chromatin remodeling are required intrinsically for self-renewal of stem cells. The equilibrium between maintenance and differentiation is regulated by mutual actions of extracellular environment and its intracellular factors. As stem cells have several key intrinsic properties that distinguish them from other cells, they may be allowed to play a major role in organizing their own niche, however their natural immune characteristics still depend on those environmental factors released from niche. Of particular interest, we need to know more about the source of these intrinsic properties and how they act upon niche signaling. Drosophila GSCs are an effective system for studying the role of intrinsic factors in stem cell maintenance. Germline stem cells are derived from PGCs. It is characterized by asymmetric localizing of specific factors known as the pole plasm [93]. Among pole plasm elements are, Nanos and Pumili .They are essential for some of the PGCs and GSCs characteristic in Drosophila, like the spherical appearance of spectrosomes during embryogenesis [10, 11]. Recent reports showed that the PGCs mainte- nance required both genes through stages of development [9, 11, 94]. It has also been shown that regulation of post- transcriptional gene plays a fundamental role in establishing and maintaining stem cell identity. The genome is transcriptionally inactive and gene expression relies on stored maternal mRNAs. During Drosophila oogenesis, the GSC identity maintenance takes place mainly by stopping the expression of factors that activate differentiation. This is can be achieved *via* activating of various genes function either in the GSC or in its niche. Pumilio and Nanos prevent differentiation, by inhibiting the translation of early differentiation factors in GSCs. These genes show an old mechanism for specification of germline used by the animal kingdom [95]. In Drosophila, Piwi, encodes the founding member of a highly conserved family of proteins Piwi, Aubergine (Aub), and Ago3. They act in concert with their small RNA guides, the Piwi-interacting RNAs (piRNAs) [96, 97]. Previous results indicated that nuclear Piwi protein plays multiple roles through transposon silencing in both germ and somatic follicle cells, [10, 98, 99]. It is essential factor for fGSC maintenance in the female germline and in bam silencing based on signals from niche [23, 100]. Piwi are predominantly present in the germline of several organisms and it is believed to act during germline development by restricting transposon expression. This may be partly through heterochromatin formation and transcriptional silencing of particular region on chromosome 3. Therefore, Piwi-family proteins not only function as regulator in germline stem cell renewal but also act as protector against such selfish DNA elements, transposon. Surprisingly, Piwi and fGSCs interactions rely mainly on single piRNA 3R- TAS1 expression [31, 101]. This activates the modifications of euchromatic histone and transcription of piRNA [101]. Therefore, PiWi may perform a control function of the epigenetic state of fGSCs through regional regulators and protein-protein interactions. Other mechanism that regulates the Drosophila GSCs maintenance in their niche is the microRNA interference pathway which is essential in the regulation of various fundamental processes [102]. Dicer-1 (Dcr- 1) and Argonaute protein, Ago1 activation is required in different stages; Dcr-1 controls fGSC division in adult germline probably by controlling the expression of the cell cycle suppressor dacapo [103, 104]. Moreover, Ago1, Dcr-1 and Loqs are independently needed for the maintenance of fGSCs for long duration [105 - 108]. Dcr- 1 is also needed for ovarian FSCs maintenance [106] suggesting that miRNA-related mechanisms my function as innate factor in maintenance of the stem cell such as Mei-P26 in fGSCs. It has the ability to interact with Ago1 to prevent miRNA expression and function during germ cell differentiation [109]. The control of nuclear organization in stem cells together with certain epigenetic regulations are considered to be important intrinsic property that influence stem cell maintenance. Hyperdynamic plasticity of chromatin proteins are innately needed for stem cell maintenance. In Drosophila's ovary, two ATP-dependent chromatin remodeling factors, limitation SWI (lSWI) and DOMINO (DOM) are needed for the self renewal of FSCs and fGSC [30]. Moreover, Stonewall (Stwl), a DNA-associated protein is essential for the maintenance of fGSCs [110]. Unlike Piwi and several previously described GSC maintenance factors, Stwl does not act on bam, It stops the expression of many genes, including genes that may be regulated by Nanos and Pumilio translational repression [106, 108, 110].

MODE OF DROSOPHILA STEM CELL DIVISION AND REGULATION OF CENTROSOME DYNAMICS

In mitotic division , the mitotic spindle shows association with the spectrosome in fGSCs [111], while in mGSCs, the mitotic spindle shows association to a closer degree with the cortex, providing a more basolateral location of spectrosome with respect to the hub.The best-studied example for an intrinsically asymmetric division is Drosophila NBs [reviewed by 112, 113]. As larval brain neuroblasts are very similar to Drosophila GSCs asymmetric behavior. They orient their spindle fibers in the direction of the final division, then divide perpendicularly to produce another (apical) NB and a smaller, basally placed ganglion mother cell (GMC) that will differentiate into neurons or glia. NB polarity is necessary for correct spindle arrangement and asymmetric localization. It is directed by recruiting several apical protein complexes including apical Par/Inscuteable, Pins,

Gαl and Mushroom body defective (Mud) [112, 113]. It is thought that the NB centrosome acts as a reference point for apical accumulation of the Par complex during interphase to establish an apicobasal axis of polarity within the cell. Both centrioles move a part. One of them is active keeping pericentriolar matter and forms a "dominant centrosome." This centrosome are the main microtubule organizing center (MTOC) in the cell and stays static, composing one polar spindle in future. The other one is initially inactive and moves to the other side of the cell at the start of mitosis [114, 115]. Indeed mutations affecting MTOC formation lead to relatively defects in NB asymmetric divisions [115]. In addition, a defect in the apical Par/Inscuteable complex elements and Pins causes centriole destabilization and leads to unusual NB divisions which influences spindle direction and cell size as well [114]. Hence, differential regulation of centrosomes appears to be an important property of asymmetric NB divisions. Research work on Drosophila testis by Yamashita *et al*. has proposed that the control role of centrosomes is fundamental in governing the asymmetric mGSC divisions [116]. Both male and female GSCs possess polarity in the context of their niche. Unlike NBs, they remain contact with their stromal cap or hub cells [69, 72, 117]. Accordingly, In male GSCs, mitosis is directed in a way that daughter cells are existing proximal and distal to the hub [117]. This depends on a particular part of the mGSC cortex which possesses adherens junctions and adenomatous polyposis coli protein homologue (APC2), E- cadherin and Armadillo (ARM) which play major roles in GSC anchoring during mitoses. Moreover, the communication of astral microtubules in the presence of APC2 seems to participate in controlling spindle alignment during mGSC division [117]. Interestingly, mutations in APC2 or centrosome components lead to aberrant mitotic spindle orientation in GSCs. This causes excess number of daughter cells contact the hub [117]. Studies in Drosophila female GSCs showed, that the differentiating cells inherit the mother centrosome [117]. Further investigations are needed to elucidate whether the inherited centrosome has a role in deciding the GSC fate. Male GSCs and NBs are thought to perform a conserved mechanism regulating asymmetric cell division. During mitosis, apical spectrosome, is essential in polarity establishment and spindle orientation [111]. Accordingly, disruption of spectrosomes by mutations leads to defect in spindle orientation and consequently the potential of GSCs asymmetric division in ovary [10, 111]. However, Stevens *et al*. suggest that the randome pattern of centrosome position is essential in fGSC during interphase, even after the centrosome split aprt [118]. More surprising, Drosophila fGSCs mutants that lack functional centrosomes normally complete the formation and orientation of a mitotic spindle along apical–basal axis [118]. This fact raises an interesting paradox and a question; is centrosome essential for asymmetric female GSC division?. Further investigations about the exact complexion of centrosome regulatory mechanism in both male and female GSCs asymmetric divisions are needed. However, the present findings propose clear differences between mGSC and fGSC that might be interpreted by the differences in spindle apparatus. Among the differences of mitotic spindle orientation and positioning in mGSC and fGSC, is the extra basolateral localization of mitotic spindle in fGSC relating to the hub [117].

Obviously it is important to define structural differences between fGSC and mGSC spectrosomes which allows them to direct spindle apparatus without the control of the centrosomes.

Stem cell niches are composed of supporting cells and associated extracellular matrix (ECM). The spatial organization of stem cells into a niche is a key factor for growth and continual tissue renewal during development, and regeneration. How possible the system of stem cells and associated neighboring cells become organized? Properly regulated cell adhesion is believed to link several major developmental signaling path ways which are crucial for cell morphogenesis. As a matter of fact, the equilibrium of cell-cell adhesion property can orchestrate both changes in arrangements and cellular behavior including mobilization, developing of cellular compositions like epithelial sheets, sorting and isolating of cell population [65]. Work largely on Drosophila has started to reveal the molecular mechanisms that participate in the maintenance of stem cell identity. Two fundamental mechanisms, cell signaling and adhesive mechanisms, govern the process. The elements in adhesion mechanism, DE-cadherin and Armadillo/β-catenin, are concentrated in the junction between GSCs and cap as junctions. The existence of this kind of junctions is commonly shown by concentration of their component s at high levels in the surfaces of the cells involved [66, 65]. Moreover, other adhesion molecules which are accumulated between GSCs and stromal cells [67 - 70], indicate their essential role in the stem cell organization. In fact, mutant fGSC clones that were lost very quickly in the disturbance of adherens junction-mediated cell adhesion confirm that proper cell adhesion between GSCs and niche cells is critical for GSCs structure, function and self-renewal mechanisms. [69, 71, 72]. Moreover, recent findings suggest that niche-germline cell adhesion that are mediated by Cadherin probably have GSCs [73, 74] and Rab11-mediated trafficking of DE-Cadherin should have a role in a competitive relationship between normal Drosophila ovarian germline and those with differentiation defects for niche occupancy. This occurs by up regulating E-cadherin expression [75]. However, a lack of expression of DE-Cadherin has been specifically related to the loss of the maintenance of stem cell and the integrity of their niche in male and female gonads during aging process [76, 77]. Adherens junctions also seem essentialfor maintaining other cell populations throughout the ovary, like the somatic FSCs [78].

Recently, various studies have focused on the mechanism of stem cells maintenance in their niches. They indicate that male GSC niche is likely to form during embryogenesis, sincea subset of anterior somatic gonadal precursors (SGPs) start to accumulate Fasciclin III, DE- and N- Cadherin adhesion proteins, similar to those of adult hub cells [79]. During the final stages of embryogenesis a small number of primordial germ cells (PGCs) PGCs are organized into a characteristic rosette-like structure [79]. Moreover, it is believed that GTPase signaling plays an important role in adhesion of niche cells. Mutation of Rap-GEF seems to be defective specifically for the recruitment of adherent junction elements between hub cells and mGSCs. Additionally germline stem cells were observed during larva development in Rap-GEF mutants. Although Rap GTPase activity seems to primarily affect the hub cells, it sloss is followed by a mechanism of compensation through over expressing of DE-Cadherin in mGSCs. This indicates that cell attachment is essential for niche function [72]. The Extracellular Matrix (ECM) is thought to be an important component of niche because in many cases, stem cells such as ISCs and FSCs, directly contact the ECM. They lack direct association with identified stromal cells and are located adjacent to the basement membrane. Thus the identification of such putative stem cell niches [80, 81] has clarified the essential role of stem cell attachment to the basement membrane and the indirect cell-to-cell communication that is mediated *via* the Extracellular Matrix (ECM). During the early stages of embryo development, ECM elements gather around the male gonad and by the late stage of embryogenesis the hub cells appear to communicate with the ECM by adopting their anterior position [82]. It has also been shown that integrin is critical for appropriate hub cells attachment to the ECM. Moreover defect in integrin lead to an improper position of embryonic hub cells during gonad morphogenesis. Recently, it has been suggested that ECM has a key role not only in niche morphology but in regulation of niche signaling as well as the IV collagen extracellular matrix proteins are essential in the formation of BMP gradient in the niche of fGSC [89]. However, how the ECM is established and maintained in those niches that lack stable stromal cell associations, a mechanism still remains to be elucidated.

DOES IT DIFFER BETWEEN MALE AND FEMALE FLIES? DROSOPHILA

Germline stem cells in male and female drosophila have been emerged as an important model system to study the communication between stem cells and their niche. Although the GSC differentiation pathways that both testis and ovary share are common, their communication in the corresponding organs is different. Gene analysis studies have determined individual signaling pathways to define permissive zone for stem cell self-renewal [83, 84]. In GSC niche in Drosophila ovaries the asymmetric cell division is likely to correlate another contributing factors regenerating from the niche. BMPs and Piwi, are produced from cap cells to regulate GSC division in ovary through intracellular signaling pathways. Dpp and Gbb function in activation of BMP together with the ot fin protein [85]. Further, JAK-STAT signaling is crucial for self- renewal of drosophila testis. Additionly, Unpaired (upd) contributes in STAT activation and expression of Zfh1 [86]. The greatly expanded studies on Drosophila stem cells and niches indicate that there are strong similarities in signaling pathways involved in both mGSC and fGSC niches regardless differences in anatomy [87].

Recently, it has been described that the JAK/STAT signalingpathway plays an important role in niche function in ovary [21, 88, 89]. Conversely, when JAK/STAT signaling is increased by driving the unpaired ligand upd in germarial somatic cells, the number of cysts increased, and the germarium became disorganized, demonstrating that this pathway is enough to retain the female germline stem cell niche in the ovary. However, the JAK/STAT pathway is not required for male stem cells maintenance in testis [21, 88], which means that it may mediate indirect effects. Upd2 promote mGSC tumor formation significantly and requires increasing in dpp expression and activation of the BMP signaling pathway [21]. In addition, JAK/STAT signaling in ESCs is probably required for their maintenance [88]. In agreement with this, the loss of hop is enough to induce GSC differentiation [21, 89]. The communication between support cells and stem cells *via* JAK/STAT and BMP signaling pathways in the fGSC niche has shed light on whether male and female GSC niches develop and operate using similar regulatory mechanisms. In fact, JAK/STAT signaling pathway is essential in maintaining mGSCs self-renewal as well [90, 91]. Furthermore, this signaling does not act alone, but in conjunction with BMP signaling to ensure the GSC renewal. A strong suppression of bam expression follow sectopic activation of the BMP signaling pathway in the germ line of male, However, this is not enough to induce the formation of mGSC tumor [90, 91]. It also appears that tumors induced by upd are unable to express bam [37]. Since signals controlled by JAK\STAT and BMP in the germ line repress bam expression, it is possible that these pathways interact in germ line cells. The association of upd-induced tumors with activation of the BMP pathway in mGSCs is yet to be reported. BMP seems to have indirect effect on gonads since their gradient ligands are not visualized. However, the regulatory role of BMP and Bam in turning from cyst divisions amplification to spermatogenesis initiation [91, 92] indicates that BMP ligands probably distributed more in the testis than in the ovary. This probably explains the reason why JAK\STAT signaling is a limiting key element involved in defining the male GSC niche.

Multiple signaling pathways have been reported to contribute to the regulation of stem cell self-renewal. However, it is clear that different molecules probably function by underlying mechanisms in various types of stem cell niches .Among these underlying mechanisms: RNA expression control and chromatin remodeling are required intrinsically for stem cell self-renewal. The equilibrium between stem cell maintenance and differentiation is regulated by mutual actions of extracellular environment and its intracellular factors. As stem cells have several key intrinsic properties that distinguish them from other cells, they may be allowed to play a major role in organizing their own niche, however their natural immune characteristics still depend on those environmental factors released from niche. Of particular interest, we need to know more about the source of these intrinsic properties and how they act upon niche signaling. Drosophila GSCs are an effective system for studying the role of intrinsic factors in stem cell maintenance. Germline stem cells are derived from PGCs. It is characterized by asymmetric localizing of specific factors known as the pole plasm, [93]. Among pole plasm elements are, Nanos and Pumili. They are essential for some of the PGCs and GSCs characteristic in Drosophila like the spherical appearance of spectrosomes during embryogenesis [10, 11]. Recent reports showed thatboth genes are required for the maintenance of PGCs through stages of development, then fGSCs in the adult ovary [9, 11, 94]. Thus, both factors piwi and Pumilio are considered to be essential for activating and/or maintaining particular gene expression programs in stem cell identity.

It has also been shown that regulation of post- transcriptional gene plays a fundamental role in establishing and maintaining stem cell identity. The genome is transcriptionally inactive and gene expression relies on stored maternal mRNAs. During Drosophila oogenesis, the GSC identity maintenance takes place mainly by stopping the expression of factors that activate differentiation. This is can be achieved *via* activating of various genes function either in the GSC or in its niche. Pumilio and Nanos prevent differentiation, by inhibiting the translation of early differentiation factors in GSCs. These genes show an old mechanism for specification of germline used by the animal kingdom [95]. In Drosophila, Piwi, encodes the founding member of a highly conserved family of proteins Piwi, Aubergine (Aub), and Ago3. They act in concert with their small RNA guides, the Piwi-interacting RNAs (piRNAs) [96, 97]. Previous results indicated that nuclear Piwi protein plays multiple roles through transposon silencing in both germ and somatic follicle cells, [10, 98, 99]. It is essential factor for fGSC maintenance in the female germline and in bam silencing based on signals from niche [23, 100]. Piwi are predominantly present in the germline of several organisms and it is believed to act during germline development by restricting transposon expression.

This may be partly through heterochromatin formation and transcriptional silencing of particular region on chromosome 3. Therefore, Piwi-family proteins not only function as regulator in germline stem cell renewal but also act as protector against such selfish DNA elements, transposon. Surprisingly, Piwi and fGSCs interactions rely mainly on single piRNA 3R- TAS1 expression [31, 101]. This activates the modifications of euchromatic histone and transcription of piRNA [123]. Therefore, PiWi may perform a control function of the epigenetic state of fGSCs through regional regulators and protein-protein interactions .

Other mechanism that regulate the Drosophila GSCs maintenance in their niche is the microRNA interference pathway. It is essential in the regulation of various fundamental processes [102] it needs Dicer-1 (Dcr- 1) and Argonaute protein, Ago1 activation. In the ovary, Dcr-1 is intrinsically control fGSC division within adult germline probably by controlling the expression of a cell cycle suppressor dacapo [103 , 104]. Moreover, Ago1, Dcr-1 and Loqs are independently needed for the maintenance of fGSCs for long duration [105 - 108]. Dcr- 1 is also needed for ovarian FSCs maintenance [106] suggesting that miRNA-related mechanisms my function as innate factor in maintenance of the stem cell such as Mei-P26 in fGSCs. It has the ability to interact with Ago1 to prevent miRNA expression and function during germ cell differentiation [109]. The control of nuclear organization in stem cells together with certain epigenetic regulations are considered to be important intrinsic property that influence stem cell maintenance. Hyperdynamic plasticity of chromatin proteins are innately needed for stem cell maintenance. In Drosophila's ovary, two ATP-dependent chromatin remodeling factors, limitation SWI (lSWI) and DOMINO (DOM) are needed for the self renewal of FSCs and fGSC [30]. Moreover, Stonewall (Stwl), a DNA-associated protein is essential for the maintenance of fGSCs [110]. Unlike Piwi and several previously described GSC maintenance factors, Stwl does not act on bam, It stops the expression of many genes, including genes that may be regulated by Nanos and Pumilio translational repression [106, 108, 110].

MODE OF DROSOPHILA STEM CELL DIVISION AND REGULATION OF CENTROSOME DYNAMICS

In fGSCs, the mitotic spindle shows association with the spectrosome [111], while in mGSCs, the mitotic spindle shows association to a closer degree with the cortex, presuming with the spectrosome a more basolateral location with respect to the hub. The best-studied example for an intrinsically asymmetric division is Drosophila NBs [reviewed by 112, 113]. Larval brain neuroblasts are very similar to Drosophila GSCs asymmetric behavior. They orient their mitotic spindle in the direction that they will finally divide. Then divide perpendicularly to produce another (apical) NB and a smaller, basally placed ganglion mother cell (GMC) that will differentiate into neurons or glia. NB polarity is necessary for correct spindle arrangement and asymmetric localization. It is directed by the apical Par/Inscuteable protein complex, which finds out spindle orientation, spindle positioning and the basal localization of cell fate determinants such as Numb and Prospero [112, 113]. Another apical accumulation of protein complexes containing Pins, Gαl and Mushroom body defective (Mud) appears to control the spindle orientation [113]. It is thought that the NB centrosome acts as a reference point for apical accumulation of the Par complex during interphase to establish an apicobasal axis of polarity within the cell. Both centrioles move a part , one of them is active keeping pericentriolar matter and forms a "dominant centrosome" This centrosome are the main microtubule organizing center (MTOC) in the cell and stays static, composing one polar spindle in future [114, 115]. The second one is initially inactive and moves to the other side of the cell at the start of mitosis [114, 115]. Indeed mutations affecting MTOC formation lead to relatively defects in NB asymmetric divisions [115]. In addition, a defect in the apical Par/Inscuteable complex elements Pins causes destabilization to both centrioles and leads to unusual NB divisions which influence spindle direction and cell size as well [114]. Hence, differential regulation of centrosomes appears to be an important property of asymmetric NB divisions. Research work on Drosophila testis by Yamashita *et al.* has proposed that differential control of the centrosomes is a key factor in governing the asymmetric mGSC divisions [116]. Both male and female GSCs possess polarity in the context of their niche. Unlike NBs, they remain contact with their stromal cap or hub cells [69, 72, 117]. Accordingly, In the male GSCs, mitosis is oriented so that daughter cells are located proximal and distal to the hub [118]. A particular part of the mGSC cortex that possesses adherens junctions and adenomatous polyposis coli protein homologue (APC2), E-cadherin and Armadillo (ARM) appears to supply a polarity cue or anchor for astral microtubules that emerge from the centrosome. The communication of astral microtubules in the presenceofAPC2 seems to participate in controlling spindle alignment during mGSC division [118]. Interestingly, mutations in APC2 or centrosome components lead to aberrant mitotic spindle orientation in GSCs. This causes excess number of daughter cells contact the hub [118].

Studies in Drosophila female GSCs showed, that the differentiating cells inherit the mother centrosome [118]. Further investigations are needed to elucidate whether the inherited centrosome by mGSCs has a role in deciding the GSC fate. Male GSCs and NBs are thought to perform a conserved mechanism regulating asymmetric cell division. During mitosis, apical spectrosome, is essential in polarity establishment and spindle orientation [111]. Accordingly, disruption of spectrosomes by mutations lead to defect in spindle orientation and consequently the potential of GSCs asymmetric division in ovary [10, 111]. However, Stevens *et al.* suggest that the randome pattern of centrosome position is essential in fGSC during interphase, even after the centrosome split apart [119]. More surprising, Drosophila fGSCs mutants that lack functional centrosomes normally complete the formation and orientation of a mitotic spindle along apical–basal axis [119].This fact raises an interesting paradox and a question; is centrosome essential for asymmetric female GSC division?. Further investigations about the exact complexion of centrosome regulatory mechanism in both male and female GSCs asymmetric divisions are needed. However, the present findings propose clear differences between mGSC and fGSC that might be interpreted by the differences in spindle apparatus. Among the differences of mitotic spindle orientation and positioning in mGSC and fGSC, is the extra basolateral localization of mitotic spindle in fGSC relating to the hub [118].

It would be important to define whether fGSC spectrosomes have a different constitution from that inmGSC spectrosome which allows them to direct spindle apparatus without the control of the centrosomes.

CONFLICT OF INTEREST

The author confirms that author has no conflict of interest to declare for this publication.

ACKNOWLEDGEMENTS

Declared none.

REFERENCES

[1]   Pepling ME, Spradling AC. Female mouse germ cells form synchronously dividing cysts. Development 1998; 125(17): 3323-8.
       [PMID: 9693136]

[2]   Pepling ME, Spradling AC. Mouse ovarian germ cell cysts undergo programmed breakdown to form primordial follicles. Dev Biol 2001; 234(2): 339-51.
       [http://dx.doi.org/10.1006/dbio.2001.0269] [PMID: 11397004]

[3]   Akashi K. Lineage promiscuity and plasticity in hematopoietic development. Ann N Y Acad Sci 2005; 1044: 125-31.
       [http://dx.doi.org/10.1196/annals.1349.016] [PMID: 15958705]

[4]   Von Stetina JR, Orr-Weaver TL. Developmental control of oocyte maturation and egg activation in metazoan models. Cold Spring Harb Perspect Biol 2011; 3(10): a005553.
       [http://dx.doi.org/10.1101/cshperspect.a005553] [PMID: 21709181]

[5]   Artavanis-Tsakonas S, Simpson P. Choosing a cell fate: a view from the Notch locus. Trends Genet 1991; 7(11-12): 403-8.
       [http://dx.doi.org/10.1016/0168-9525(91)90264-Q] [PMID: 1668193]

[6]   Austin J, Kimble J. glp-1 is required in the germ line for regulation of the decision between mitosis and meiosis in *C. elegans*. Cell 1987; 51(4): 589-99.
       [http://dx.doi.org/10.1016/0092-8674(87)90128-0] [PMID: 3677168]

[7]   Cinquin O, Crittenden SL, Morgan DE, Kimble J. Progression from a stem cell-like state to early differentiation in the *C. elegans* germ line. Proc Natl Acad Sci USA 2010; 107(5): 2048-53.
       [http://dx.doi.org/10.1073/pnas.0912704107] [PMID: 20080700]

[8]   Gilboa L, Lehmann R. Soma-germline interactions coordinate homeostasis and growth in the Drosophila gonad. Nature 2006; 443(7107): 97-100.
       [http://dx.doi.org/10.1038/nature05068] [PMID: 16936717]

[9]   Gilboa L, Lehmann R. Repression of primordial germ cell differentiation parallels germ line stem cell maintenance. Curr Biol 2004; 14(11): 981-6.
       [http://dx.doi.org/10.1016/j.cub.2004.05.049] [PMID: 15182671]

[10]   Lin H, Spradling AC. A novel group of pumilio mutations affects the asymmetric division of germline stem cells in the Drosophila ovary. Development 1997; 124(12): 2463-76.
       [PMID: 9199372]

[11]   Forbes A, Lehmann R. Nanos and Pumilio have critical roles in the development and function of Drosophila germline stem cells. Development 1998; 125(4): 679-90.
       [PMID: 9435288]

[12]   Wang Z, Lin H. Nanos maintains germline stem cell self-renewal by preventing differentiation. Science 2004; 303(5666): 2016-9.

[http://dx.doi.org/10.1126/science.1093983] [PMID: 14976263]

[13]   Fichelson P, Moch C, Ivanovitch K, *et al.* Live-imaging of single stem cells within their niche reveals that a U3snoRNP component segregates asymmetrically and is required for self-renewal in Drosophila. Nat Cell Biol 2009; 11(6): 685-93.
[http://dx.doi.org/10.1038/ncb1874] [PMID: 19430468]

[14]   Liu N, Han H, Lasko P. Vasa promotes Drosophila germline stem cell differentiation by activating mei-P26 translation by directly interacting with a (U)-rich motif in its 3 ? UTR. Genes Dev 2009; 23(23): 2742-52.
[http://dx.doi.org/10.1101/gad.1820709] [PMID: 19952109]

[15]   Kugler JM, Lasko P. Localization, anchoring and translational control of oskar, gurken, bicoid and nanos mRNA during Drosophila oogenesis. Fly (Austin) 2009; 3(1): 15-28.
[http://dx.doi.org/10.4161/fly.3.1.7751] [PMID: 19182536]

[16]   Klattenhoff C, Theurkauf W. Biogenesis and germline functions of piRNAs. Development 2008; 135(1): 3-9.
[http://dx.doi.org/10.1242/dev.006486] [PMID: 18032451]

[17]   Kotaja N, Sassone-Corsi P. The chromatoid body: a germ-cell-specific RNA-processing centre. Nat Rev Mol Cell Biol 2007; 8(1): 85-90.
[http://dx.doi.org/10.1038/nrm2081] [PMID: 17183363]

[18]   Lin MD, Jiao X, Grima D, Newbury SF, Kiledjian M, Chou TB. Drosophila processing bodies in oogenesis. Dev Biol 2008; 322(2): 276-88.
[http://dx.doi.org/10.1016/j.ydbio.2008.07.033] [PMID: 18708044]

[19]   Soper SF, van der Heijden GW, Hardiman TC, *et al.* Mouse maelstrom, a component of nuage, is essential for spermatogenesis and transposon repression in meiosis. Dev Cell 2008; 15(2): 285-97.
[http://dx.doi.org/10.1016/j.devcel.2008.05.015] [PMID: 18694567]

[20]   Voronina E, Seydoux G, Sassone-Corsi P, Nagamori I. RNA granules in germ cells. Cold Spring Harb Perspect Biol 2011; 3(12): a002774.
[http://dx.doi.org/10.1101/cshperspect.a002774] [PMID: 21768607]

[21]   LA3pez-Onieva L, FernA ndez-MiA A n A, GonzA lez-Reyes A. Jak/Stat signalling in niche support cells regulates dpp transcription to control germline stem cell maintenance in the Drosophila ovary. Development 2008; 135(3): 533-40.
[http://dx.doi.org/10.1242/dev.016121] [PMID: 18171682]

[22]   Chen D, McKearin D. Dpp signaling silences bam transcription directly to establish asymmetric divisions of germline stem cells. Curr Biol 2003; 13(20): 1786-91.
[http://dx.doi.org/10.1016/j.cub.2003.09.033] [PMID: 14561403]

[23]   Chen D, McKearin D. Gene circuitry controlling a stem cell niche. Curr Biol 2005; 15(2): 179-84.
[http://dx.doi.org/10.1016/j.cub.2005.01.004] [PMID: 15668176]

[24]   Kai T, Spradling A. Differentiating germ cells can revert into functional stem cells in Drosophila melanogaster ovaries. Nature 2004; 428(6982): 564-9.
[http://dx.doi.org/10.1038/nature02436] [PMID: 15024390]

[25]   Civetta A, Rajakumar SA, Brouwers B, Bacik JP. Rapid evolution and gene-specific patterns of

selection for three genes of spermatogenesis in Drosophila. Mol Biol Evol 2006; 23(3): 655-62.
[http://dx.doi.org/10.1093/molbev/msj074] [PMID: 16357040]

[26]   Guo Z, Wang Z. The glypican Dally is required in the niche for the maintenance of germline stem cells and short-range BMP signaling in the Drosophila ovary. Development 2009; 136(21): 3627-35.
[http://dx.doi.org/10.1242/dev.036939] [PMID: 19793889]

[27]   Hayashi Y, Kobayashi S, Nakato H. Drosophila glypicans regulate the germline stem cell niche. J Cell Biol 2009; 187(4): 473-80.
[http://dx.doi.org/10.1083/jcb.200904118] [PMID: 19948496]

[28]   Schulz C, Wood CG, Jones DL, Tazuke SI, Fuller MT. Signaling from germ cells mediated by the rhomboid homolog stet organizes encapsulation by somatic support cells. Development 2002; 129(19): 4523-34.
[PMID: 12223409]

[29]   Casanueva MO, Ferguson EL. Germline stem cell number in the Drosophila ovary is regulated by redundant mechanisms that control Dpp signaling. Development 2004; 131(9): 1881-90.
[http://dx.doi.org/10.1242/dev.01076] [PMID: 15105369]

[30]   Xi R, Xie T. Stem cell self-renewal controlled by chromatin remodeling factors. Science 2005; 310(5753): 1487-9.
[http://dx.doi.org/10.1126/science.1120140] [PMID: 16322456]

[31]   Smulders-Srinivasan TK, Lin H. Screens for piwi suppressors in Drosophila identify dosage-dependent regulators of germline stem cell division. Genetics 2003; 165(4): 1971-91.
[PMID: 14704180]

[32]   Reichert H. Drosophila neural stem cells: cell cycle control of self-renewal, differentiation, and termination in brain development. Results Probl Cell Differ 2011; 53: 529-46.
[http://dx.doi.org/10.1007/978-3-642-19065-0_21] [PMID: 21630158]

[33]   Green P, Hartenstein AY, Hartenstein V. The embryonic development of the Drosophila visual system. Cell Tissue Res 1993; 273(3): 583-98.
[http://dx.doi.org/10.1007/BF00333712] [PMID: 8402833]

[34]   Prokop A, Technau GM. The origin of postembryonic neuroblasts in the ventral nerve cord of Drosophila melanogaster. Development 1991; 111(1): 79-88.
[PMID: 1901786]

[35]   White K, Grether ME, Abrams JM, Young L, Farrell K, Steller H. Genetic control of programmed cell death in Drosophila. Science 1994; 264(5159): 677-83.
[http://dx.doi.org/10.1126/science.8171319] [PMID: 8171319]

[36]   Maurange C, Cheng L, Gould AP. Temporal transcription factors and their targets schedule the end of neural proliferation in Drosophila. Cell 2008; 133(5): 891-902.
[http://dx.doi.org/10.1016/j.cell.2008.03.034] [PMID: 18510932]

[37]   Tulina N, Matunis E. Control of stem cell self-renewal in Drosophila spermatogenesis by JAK-STAT signaling. Science 2001; 294(5551): 2546-9.
[http://dx.doi.org/10.1126/science.1066700] [PMID: 11752575]

[38]   White K, Kankel DR. Patterns of cell division and cell movement in the formation of the imaginal

nervous system in Drosophila melanogaster. Dev Biol 1978; 65(2): 296-321.
[http://dx.doi.org/10.1016/0012-1606(78)90029-5] [PMID: 98369]

[39] Egger B, Boone JQ, Stevens NR, Brand AH, Doe CQ. Regulation of spindle orientation and neural stem cell fate in the Drosophila optic lobe. Neural Dev 2007; 2: 1.
[http://dx.doi.org/10.1186/1749-8104-2-1] [PMID: 17207270]

[40] Egger B, Gold KS, Brand AH. Notch regulates the switch from symmetric to asymmetric neural stem cell division in the Drosophila optic lobe. Development 2010; 137(18): 2981-7.
[http://dx.doi.org/10.1242/dev.051250] [PMID: 20685734]

[41] Eaves C, Glimm H, Eisterer W, Audet J, Maguer-Satta V, Piret J. Characterization of human hematopoietic cells with short-lived *in vivo* repopulating activity. Ann N Y Acad Sci 2001; 938: 63-70.
[http://dx.doi.org/10.1111/j.1749-6632.2001.tb03575.x] [PMID: 11458527]

[42] North TE, de Bruijn MF, Stacy T, *et al.* Runx1 expression marks long-term repopulating hematopoietic stem cells in the midgestation mouse embryo. Immunity 2002; 16(5): 661-72.
[http://dx.doi.org/10.1016/S1074-7613(02)00296-0] [PMID: 12049718]

[43] Growney JD, Shigematsu H, Li Z, *et al.* Loss of Runx1 perturbs adult hematopoiesis and is associated with a myeloproliferative phenotype. Blood 2005; 106(2): 494-504.
[http://dx.doi.org/10.1182/blood-2004-08-3280] [PMID: 15784726]

[44] North TE, Stacy T, Matheny CJ, Speck NA, de Bruijn MF. Runx1 is expressed in adult mouse hematopoietic stem cells and differentiating myeloid and lymphoid cells, but not in maturing erythroid cells. Stem Cells 2004; 22(2): 158-68.
[http://dx.doi.org/10.1634/stemcells.22-2-158] [PMID: 14990855]

[45] Evans C, Sinenko S, Mandal L, Martinez-Agosto J, Hartenstein V, Banerjee U. Genetic dissection of hematopoiesis using *Drosophila* as a model system. In: Wassarman P, Ed. Advances in developmental biology. New York: Elsevier Science 2007; pp. 259-99.
[http://dx.doi.org/10.1016/S1574-3349(07)18011-X]

[46] Cherry S, Silverman N. Host-pathogen interactions in *drosophila*: new tricks from an old friend. Nat Immunol 2006; 7(9): 911-7.
[http://dx.doi.org/10.1038/ni1388] [PMID: 16924255]

[47] Jiravanichpaisal P, Lee BL, Soderhall K. Cell-mediated immunity in arthropods: hematopoiesis, coagulation, melanization and opsonization. Immunobiology 2006; 211(4): 213-36.
[http://dx.doi.org/10.1016/j.imbio.2005.10.015] [PMID: 16697916]

[48] Lemaitre B, Hoffmann J. The host defense of *Drosophila melanogaster.* Annu Rev Immunol 2007; 25: 697-743.
[http://dx.doi.org/10.1146/annurev.immunol.25.022106.141615] [PMID: 17201680]

[49] Williams MJ. *Drosophila* hemopoiesis and cellular immunity. J Immunol 2007; 178(8): 4711-6.
[http://dx.doi.org/10.4049/jimmunol.178.8.4711] [PMID: 17404248]

[50] Evans CJ, Banerjee U. Transcriptional regulation of hematopoiesis in Drosophila. Blood Cells Mol Dis 2003; 30(2): 223-8.
[http://dx.doi.org/10.1016/S1079-9796(03)00028-7] [PMID: 12732186]

[51] Evans CJ, Hartenstein V, Banerjee U. Thicker than blood: conserved mechanisms in Drosophila and

vertebrate hematopoiesis. Dev Cell 2003; 5(5): 673-90.
[http://dx.doi.org/10.1016/S1534-5807(03)00335-6] [PMID: 14602069]

[52]　Cumano A, Godin I. Ontogeny of the hematopoietic system. Annu Rev Immunol 2007; 25: 745-85.
[http://dx.doi.org/10.1146/annurev.immunol.25.022106.141538] [PMID: 17201678]

[53]　de Jong JL, Zon LI. Use of the zebrafish system to study primitive and definitive hematopoiesis. Annu Rev Genet 2005; 39: 481-501.
[http://dx.doi.org/10.1146/annurev.genet.39.073003.095931] [PMID: 16285869]

[54]　Spradling A, Fuller MT, Braun RE, Yoshida S. Germline stem cells. Cold Spring Harb Perspect Biol 2011; 3(11): a002642.
[http://dx.doi.org/10.1101/cshperspect.a002642.] [PMID: 21791699]

[55]　Tepass U, Fessler LI, Aziz A, Hartenstein V. Embryonic origin of hemocytes and their relationship to cell death in *Drosophila*. Development 1994; 120(7): 1829-37.
[PMID: 7924990]

[56]　Bataille L, Auge B, Ferjoux G, Haenlin M, Waltzer L. Resolving embryonic blood cell fate choice in *Drosophila*: interplay of GCM and RUNX factors. Development 2005; 132(20): 4635-44.
[http://dx.doi.org/10.1242/dev.02034] [PMID: 16176949]

[57]　Lebestky T, Chang T, Hartenstein V, Banerjee U. Specification of Drosophila hematopoietic lineage by conserved transcription factors. Science 2000; 288(5463): 146-9.
[http://dx.doi.org/10.1126/science.288.5463.146] [PMID: 10753120]

[58]　Chan WY, Kohsaka S, Rezaie P. The origin and cell lineage of microglia: new concepts. Brain Res Brain Res Rev 2007; 53(2): 344-54.
[http://dx.doi.org/10.1016/j.brainresrev.2006.11.002] [PMID: 17188751]

[59]　Palis J, Robertson S, Kennedy M, Wall C, Keller G. Development of erythroid and myeloid progenitors in the yolk sac and embryo proper of the mouse. Development 1999; 126(22): 5073-84.
[PMID: 10529424]

[60]　Bertrand JY, Jalil A, Klaine M, Jung S, Cumano A, Godin I. Three pathways to mature macrophages in the early mouse yolk sac. Blood 2005; 106(9): 3004-11.
[http://dx.doi.org/10.1182/blood-2005-02-0461] [PMID: 16020514]

[61]　Tober J, Koniski A, McGrath KE, *et al.* The megakaryocyte lineage originates from hemangioblast precursors and is an integral component both of primitive and of definitive hematopoiesis. Blood 2007; 109(4): 1433-41.
[http://dx.doi.org/10.1182/blood-2006-06-031898] [PMID: 17062726]

[62]　Mikkola HK, Orkin SH. The journey of developing hematopoietic stem cells. Development 2006; 133(19): 3733-44.
[http://dx.doi.org/10.1242/dev.02568] [PMID: 16968814]

[63]　Samokhvalov IM, Samokhvalova NI, Nishikawa S. Cell tracing shows the contribution of the yolk sac to adult haematopoiesis. Nature 2007; 446(7139): 1056-61.
[http://dx.doi.org/10.1038/nature05725] [PMID: 17377529]

[64]　Huber TL, Kouskoff V, Fehling HJ, Palis J, Keller G. Haemangioblast commitment is initiated in the primitive streak of the mouse embryo. Nature 2004; 432(7017): 625-30.

[http://dx.doi.org/10.1038/nature03122] [PMID: 15577911]

[65]   McNeill H. Sticking together and sorting things out: adhesion as a force in development. Nat Rev Genet 2000; 1(2): 100-8.
[http://dx.doi.org/10.1038/35038540] [PMID: 11253649]

[66]   Halbleib JM, Nelson WJ. Cadherins in development: cell adhesion, sorting, and tissue morphogenesis. Genes Dev 2006; 20(23): 3199-214.
[http://dx.doi.org/10.1101/gad.1486806] [PMID: 17158740]

[67]   Jenkins AB, McCaffery JM, Van Doren M. Drosophila E-cadherin is essential for proper germ cell-soma interaction during gonad morphogenesis. Development 2003; 130(18): 4417-26.
[http://dx.doi.org/10.1242/dev.00639] [PMID: 12900457]

[68]   Kiger AA, White-Cooper H, Fuller MT. Somatic support cells restrict germline stem cell self-renewal and promote differentiation. Nature 2000; 407(6805): 750-4.
[http://dx.doi.org/10.1038/35037606] [PMID: 11048722]

[69]   Song X, Zhu CH, Doan C, Xie T. Germline stem cells anchored by adherens junctions in the Drosophila ovary niches. Science 2002; 296(5574): 1855-7.
[http://dx.doi.org/10.1126/science.1069871] [PMID: 12052957]

[70]   Tazuke SI, Schulz C, Gilboa L, *et al.* A germline-specific gap junction protein required for survival of differentiating early germ cells. Development 2002; 129(10): 2529-39.
[PMID: 11973283]

[71]   GonzA lez-Reyes A. Stem cells, niches and cadherins: a view from Drosophila. J Cell Sci 2003; 116(Pt 6): 949-54.
[http://dx.doi.org/10.1242/jcs.00310] [PMID: 12584239]

[72]   Wang H, Singh SR, Zheng Z, *et al.* Rap-GEF signaling controls stem cell anchoring to their niche through regulating DE-cadherin-mediated cell adhesion in the Drosophila testis. Dev Cell 2006; 10(1): 117-26.
[http://dx.doi.org/10.1016/j.devcel.2005.11.004] [PMID: 16399083]

[73]   Boyle M, Wong C, Rocha M, Jones DL. Decline in self-renewal factors contributes to aging of the stem cell niche in the Drosophila testis. Cell Stem Cell 2007; 1(4): 470-8.
[http://dx.doi.org/10.1016/j.stem.2007.08.002] [PMID: 18371382]

[74]   Pan L, Chen S, Weng C, *et al.* Stem cell aging is controlled both intrinsically and extrinsically in the Drosophila ovary. Cell Stem Cell 2007; 1(4): 458-69.
[http://dx.doi.org/10.1016/j.stem.2007.09.010] [PMID: 18371381]

[75]   Jin Z, Kirilly D, Weng C, *et al.* Differentiation-defective stem cells outcompete normal stem cells for niche occupancy in the Drosophila ovary. Cell Stem Cell 2008; 2(1): 39-49.
[http://dx.doi.org/10.1016/j.stem.2007.10.021] [PMID: 18371420]

[76]   Bogard N, Lan L, Xu J, Cohen RS. Rab11 maintains connections between germline stem cells and niche cells in the Drosophila ovary. Development 2007; 134(19): 3413-8.
[http://dx.doi.org/10.1242/dev.008466] [PMID: 17715175]

[77]   Lighthouse DV, Buszczak M, Spradling AC. New components of the Drosophila fusome suggest it plays novel roles in signaling and transport. Dev Biol 2008; 317(1): 59-71.

[http://dx.doi.org/10.1016/j.ydbio.2008.02.009] [PMID: 18355804]

[78]   Song X, Xie T. DE-cadherin-mediated cell adhesion is essential for maintaining somatic stem cells in the Drosophila ovary. Proc Natl Acad Sci USA 2002; 99(23): 14813-8.
[http://dx.doi.org/10.1073/pnas.232389399] [PMID: 12393817]

[79]   Le Bras S, Van Doren M. Development of the male germline stem cell niche in Drosophila. Dev Biol 2006; 294(1): 92-103.
[http://dx.doi.org/10.1016/j.ydbio.2006.02.030] [PMID: 16566915]

[80]   Nystul T, Spradling A. An epithelial niche in the Drosophila ovary undergoes long-range stem cell replacement. Cell Stem Cell 2007; 1(3): 277-85.
[http://dx.doi.org/10.1016/j.stem.2007.07.009] [PMID: 18371362]

[81]   Ohlstein B, Spradling A. The adult Drosophila posterior midgut is maintained by pluripotent stem cells. Nature 2006; 439(7075): 470-4.
[http://dx.doi.org/10.1038/nature04333] [PMID: 16340960]

[82]   Tanentzapf G, Devenport D, Godt D, Brown NH. Integrin-dependent anchoring of a stem-cell niche. Nat Cell Biol 2007; 9(12): 1413-8.
[http://dx.doi.org/10.1038/ncb1660] [PMID: 17982446]

[83]   Kirilly D, Xie T. The Drosophila ovary: an active stem cell community. Cell Res 2007; 17(1): 15-25.
[http://dx.doi.org/10.1038/sj.cr.7310123] [PMID: 17199109]

[84]   Wong MD, Jin Z, Xie T. Molecular mechanisms of germline stem cell regulation. Annu Rev Genet 2005; 39: 173-95.
[http://dx.doi.org/10.1146/annurev.genet.39.073003.105855] [PMID: 16285857]

[85]   Jiang X, Xia L, Chen D, *et al.* Otefin, a nuclear membrane protein, determines the fate of germline stem cells in Drosophila via interaction with Smad complexes. Dev Cell 2008; 14(4): 494-506.
[http://dx.doi.org/10.1016/j.devcel.2008.02.018] [PMID: 18410727]

[86]   Leatherman JL, Dinardo S. Zfh-1 controls somatic stem cell self-renewal in the Drosophila testis and nonautonomously influences germline stem cell self-renewal. Cell Stem Cell 2008; 3(1): 44-54.
[http://dx.doi.org/10.1016/j.stem.2008.05.001] [PMID: 18593558]

[87]   Fuller MT, Spradling AC. Male and female Drosophila germline stem cells: two versions of immortality. Science 2007; 316(5823): 402-4.
[http://dx.doi.org/10.1126/science.1140861] [PMID: 17446390]

[88]   Decotto E, Spradling AC. The Drosophila ovarian and testis stem cell niches: similar somatic stem cells and signals. Dev Cell 2005; 9(4): 501-10.
[http://dx.doi.org/10.1016/j.devcel.2005.08.012] [PMID: 16198292]

[89]   Wang X, Harris RE, Bayston LJ, Ashe HL. Type IV collagens regulate BMP signalling in Drosophila. Nature 2008; 455(7209): 72-7.
[http://dx.doi.org/10.1038/nature07214] [PMID: 18701888]

[90]   Kawase E, Wong MD, Ding BC, Xie T. Gbb/Bmp signaling is essential for maintaining germline stem cells and for repressing bam transcription in the Drosophila testis. Development 2004; 131(6): 1365-75.
[http://dx.doi.org/10.1242/dev.01025] [PMID: 14973292]

[91]   Shivdasani AA, Ingham PW. Regulation of stem cell maintenance and transit amplifying cell proliferation by tgf-I signaling in Drosophila spermatogenesis. Curr Biol 2003; 13(23): 2065-72.
[http://dx.doi.org/10.1016/j.cub.2003.10.063] [PMID: 14653996]

[92]   Matunis E, Tran J, Gonczy P, Caldwell K, DiNardo S. punt and schnurri regulate a somatically derived signal that restricts proliferation of committed progenitors in the germline. Development 1997; 124(21): 4383-91.
[PMID: 9334286]

[93]   Okada M. Germline cell formation in Drosophila embryogenesis. Genes Genet Syst 1998; 73(1): 1-8.
[http://dx.doi.org/10.1266/ggs.73.1] [PMID: 9546203]

[94]   Truman JW, Bate M. Spatial and temporal patterns of neurogenesis in the central nervous system of Drosophila melanogaster. Dev Biol 1988; 125(1): 145-57.
[http://dx.doi.org/10.1016/0012-1606(88)90067-X] [PMID: 3119399]

[95]   Parisi M, Lin H. Translational repression: a duet of Nanos and Pumilio. Curr Biol 2000; 10(2): R81-3.
[http://dx.doi.org/10.1016/S0960-9822(00)00283-9] [PMID: 10662662]

[96]   Lin H. piRNAs in the germ line. Science 2007; 316(5823): 397.
[http://dx.doi.org/10.1126/science.1137543] [PMID: 17446387]

[97]   O'Donnell KA, Boeke JD. Mighty Piwis defend the germline against genome intruders. Cell 2007; 129(1): 37-44.
[http://dx.doi.org/10.1016/j.cell.2007.03.028] [PMID: 17418784]

[98]   Cox DN, Chao A, Baker J, Chang L, Qiao D, Lin H. A novel class of evolutionarily conserved genes defined by piwi are essential for stem cell self-renewal. Genes Dev 1998; 12(23): 3715-27.
[http://dx.doi.org/10.1101/gad.12.23.3715] [PMID: 9851978]

[99]   Megosh HB, Cox DN, Campbell C, Lin H. The role of piwi and the mirna machinery in drosophila germline determination. Curr Biol 2006; 16(19): 1884-94.
[http://dx.doi.org/http://dx.doi.org/10.1016/j.cub.2006.08.051] [PMID: 16949822]

[100]   Szakmary A, Cox DN, Wang Z, Lin H. Regulatory relationship among piwi, pumilio, and bag-o--marbles in Drosophila germline stem cell self-renewal and differentiation. Curr Biol 2005; 15(2): 171-8.
[http://dx.doi.org/10.1016/j.cub.2005.01.005] [PMID: 15668175]

[101]   Yin H, Lin H. An epigenetic activation role of Piwi and a Piwi-associated piRNA in Drosophila melanogaster. Nature 2007; 450(7167): 304-8.
[http://dx.doi.org/10.1038/nature06263] [PMID: 17952056]

[102]   Alvarez-Garcia I, Miska EA. MicroRNA functions in animal development and human disease. Development 2005; 132(21): 4653-62.
[http://dx.doi.org/10.1242/dev.02073] [PMID: 16224045]

[103]   Hatfield SD, Shcherbata HR, Fischer KA, Nakahara K, Carthew RW, Ruohola-Baker H. Stem cell division is regulated by the microRNA pathway. Nature 2005; 435(7044): 974-8.
[http://dx.doi.org/10.1038/nature03816] [PMID: 15944714]

[104]   Shcherbata HR, Ward EJ, Fischer KA, *et al.* Stage-specific differences in the requirements for germline stem cell maintenance in the Drosophila ovary. Cell Stem Cell 2007; 1(6): 698-709.

[http://dx.doi.org/10.1016/j.stem.2007.11.007] [PMID: 18213359]

[105]   Forstemann K, Tomari Y, Du T, *et al*. Normal microRNA maturation and germ-line stem cell maintenance requires Loquacious, a double-stranded RNA-binding domain protein. PLoS Biol 2005; 3(7): e236.
[http://dx.doi.org/10.1371/journal.pbio.0030236] [PMID: 15918770]

[106]   Jin Z, Xie T. Dcr-1 maintains drosophila ovarian stem cells 2007.
[http://dx.doi.org/10.1016/j.cub.2007.01.050]

[107]   Park JK, Liu X, Strauss TJ, McKearin DM, Liu Q. The miRNA pathway intrinsically controls self-renewal of Drosophila germline stem cells. Curr Biol 2007; 17(6): 533-8.
[http://dx.doi.org/17320391]

[108]   Yang L, Chen D, Duan R, *et al*. Argonaute 1 regulates the fate of germline stem cells in Drosophila. Development 2007; 134(23): 4265-72.
[http://dx.doi.org/10.1242/dev.009159] [PMID: 17993467]

[109]   NeumA1/4ller RA, Betschinger J, Fischer A, *et al*. Mei-P26 regulates microRNAs and cell growth in the Drosophila ovarian stem cell lineage. Nature 2008; 454(7201): 241-5.
[http://dx.doi.org/10.1038/nature07014] [PMID: 18528333]

[110]   Maines JZ, Park JK, Williams M, McKearin DM. Stonewalling Drosophila stem cell differentiation by epigenetic controls. Development 2007; 134(8): 1471-9.
[http://dx.doi.org/10.1242/dev.02810] [PMID: 17344229]

[111]   Deng W, Lin H. Spectrosomes and fusomes anchor mitotic spindles during asymmetric germ cell divisions and facilitate the formation of a polarized microtubule array for oocyte specification in Drosophila. Dev Biol 1997; 189(1): 79-94.
[http://dx.doi.org/10.1006/dbio.1997.8669] [PMID: 9281339]

[112]   Wodarz A. Molecular control of cell polarity and asymmetric cell division in Drosophila neuroblasts. Curr Opin Cell Biol 2005; 17(5): 475-81.
[http://dx.doi.org/10.1016/j.ceb.2005.08.005] [PMID: 16099639]

[113]   Yu F, Kuo CT, Jan YN. Drosophila neuroblast asymmetric cell division: recent advances and implications for stem cell biology. Neuron 2006; 51(1): 13-20.
[http://dx.doi.org/10.1016/j.neuron.2006.06.016] [PMID: 16815328]

[114]   Rebollo E, Sampaio P, Januschke J, Llamazares S, Varmark H, GonzA lez C. Functionally unequal centrosomes drive spindle orientation in asymmetrically dividing Drosophila neural stem cells. Dev Cell 2007; 12(3): 467-74.
[http://dx.doi.org/10.1016/j.devcel.2007.01.021] [PMID: 17336911]

[115]   Rusan NM, Peifer M. A role for a novel centrosome cycle in asymmetric cell division. J Cell Biol 2007; 177(1): 13-20.
[http://dx.doi.org/10.1083/jcb.200612140] [PMID: 17403931]

[116]   Yamashita YM, Fuller MT. Asymmetric centrosome behavior and the mechanisms of stem cell division. J Cell Biol 2008; 180(2): 261-6.
[http://dx.doi.org/10.1083/jcb.200707083] [PMID: 18209101]

[117]   Yamashita YM, Mahowald AP, Perlin JR, Fuller MT. Asymmetric inheritance of mother *versus*

daughter centrosome in stem cell division. Science 2007; 315(5811): 518-21.
[http://dx.doi.org/10.1126/science.1134910] [PMID: 17255513]

[118] Yamashita YM, Jones DL, Fuller MT. Orientation of asymmetric stem cell division by the APC tumor suppressor and centrosome. Science 2003; 301(5639): 1547-50.
[http://dx.doi.org/10.1126/science.1087795] [PMID: 12970569]

[119] Stevens NR, Raposo AA, Basto R, St Johnston D, Raff JW. From stem cell to embryo without centrioles. Curr Biol 2007; 17(17): 1498-503.
[http://dx.doi.org/10.1016/j.cub.2007.07.060] [PMID: 17716897]

CHAPTER 5

Behavior and Asymmetric Cell Divisions of Stem Cells

Sameh Elshahawy#, Ahmed R. N. Ibrahim#, Salaheldin S. Soliman, Marwa E. Elgayyar and Ahmed El-Hashash*

Stem Cells, Regenerative Medicine and Developmental Biology Program, Children's' Hospital Los Angeles, Keck School of Medicine and Ostrow School of Dentistry, University of Southern California, 4661 Sunset Boulevard, Los Angeles, CA 90027, USA

Abstract: Stem cells are unique, rare cell types that exist within many various life forms; they have been identified in both the plant and the animal kingdoms. These cells possess two distinguishing characteristics: the capacity for self-renewal in order to preserve the stem cell pool, and pluripotency, in which they differentiate into specialized cells when particular signals are given [1]. Due to these defining qualities, stem cells have been found to be primordial players during development, tissue repair and regeneration after injury, and healthy homeostatic cell turnover. They are, therefore, a crucial driving force for fast-expanding fields of regenerative medicine and functional tissue engineering [2]. The substantial building blocks of life are embryonic stem cells (ESCs). During early embryogenesis, ESCs that have their origin in the developing blastocyst's inner cell mass (ICM) contain the capacity for pluripotency. Thus, they have the ability to become any type of cell that is required to form an entire organism. Adult stem cells are another type of stem cells that are uncommon tissue-resident cells; they are essential for the establishment, maintenance, and repair as well as regeneration of highly specialized tissues in multicellular organisms [3].

Keywords: Asymmetrical cell division, Cell spindle, Cell polarity cell fate determinants, eya1, Notch signaling, Stem cells, Symmetrical cell division.

* **Correspondence author Ahmed El-Hashash:** Developmental Biology, Regenerative Medicine and Stem Cell Program, Saban Research Institute, Children's Hospital Los Angeles, 4661 Sunset Boulevard MS 35, Los Angeles, California 90027, USA; Tel: 323-361-2764, 323-361-2258; Fax: 323-361-3613; E-mail: aelhashash@chla.usc.edu
# Equal first author

1. INTRODUCTION

All stem cells that belong to different developmental stages and organs share the capacity for self-regeneration and differentiation into certain lineages. However, they differ somewhat with regard to their developmental potential. For example, totipotent mammalian stem cells are found only in early embryos and have the capability to form complete organisms. Another type, pluripotent stem cells, resides in embryoblast, also known as the inner cell mass (ICM) of a blastocyst, and are found to have a capacity to become any type of cell in the adult organism. Recently, several tissue-specific adult stem cells were recognized and isolated, including epithelial stem cells; mesenchymal stem cells (MSCs) hematopoietic stem cells (HSCs), muscle stem cells (satellite cells), intestinal stem cells (ISCs), and germ line stem cells (GSCs). These tissue-specific adult stem cells, as well as stem cells derived from cord blood and fetal tissues may be multipotent, oligopotent, or even unipotent. These cells are essential for tissue formation, maturation and repair/regeneration as well as rejuvenation. Moreover, recent studies have provided evidence of successful origins of pluripotent stem cells in adult somatic cells by induced expression of reprogramming transcriptional factors [4].

2. SOURCES, BEHAVIOR AND CELLULAR PROPERTIES OF STEM CELLS

2.1. Stem Cell Sources

Based on their sources, stem cells are frequently categorized into three groups: Embryonic stem cells (ESCs), adult stem cells (ASCs) and induced pluripotent stem cells (iPSs). ESCs are characterized by their high pluripotent differentiation potential and telomerase activity. These cells can be isolated by *in vitro* fertilization of pre-implantation blastocysts. If left to divide, the blastocyst will give a whole embryo in the proper environment, hence the name ESCs [5].

ASCs that exist in different tissues all over the body are specialized stem cells that act as reservoir cells. Once a somatic cell loses its function, the ASC replaces it to maintain organ function, regenerates tissue and maintain tissue homeostasis. Loss of ASC in amount or activity could, consequently, result in impaired organ and tissue function, which is a distinguishing quality of the aging process. ASCs are multipotent progenitor cells that have the potential to be isolated from different types of tissues in adults such as skin, muscle, bone, bone marrow, and blood vessels [6]. They are able to become different types of cells, distinctive of the tissues that they inhabit [5]. Induced pluripotent cells (iPSs) are the newest members of the stem cells. Through cell reprogramming, we are now able to regress the adult stem cell to an earlier, dedifferentiated and more potent stage that resembles the embryonic stem cell (ESCs).

2.2. Behavior and Cellular Properties of Stem Cells: Cell Division, Cell Cycle, and Life Span

Normal tissue morphogenesis and regeneration requires maintenance of an appropriate balance between self-renewal and differentiation of tissue -specific progenitors (ASCs). Therefore, understanding behavior and cellular properties, including cell polarity, cell proliferation and differentiation, cell cycle, mode of cell division (symmetric *versus* asymmetric) and life span of tissue-specific stem/progenitor cells could result in innovative explanations for reestablishing ordinary tissue morphogenesis and/or regeneration when needed.

Cell polarity is determined by asymmetrical dispersal of cellular organelles in a solitary cell. This quality is imperative for different cellular activities, such as cell specification, relocation and asymmetric division. Cell polarity maintains a pivotal role in aiding the organization and integration of intricate molecular signals in order to allow cells to make decisions regarding fate, positioning, production, differentiation, and interaction [7, 8]. Recent investigations from this research laboratory on lung stem/ progenitor cells, for example, have demonstrated that these cells are polarized and exceedingly mitotic with distinctive perpendicular cell divisions (described below). Disruption of lung stem/ progenitor cells' ability to polarize leads to a decline of the equilibrium between self-renewal and differentiation of lung stem cells *in vivo* as well as in culture [9].

The maintenance and proliferation of stem/progenitor cells relies on the method of cell division.

Stem/progenitor cells can participate in symmetric and asymmetric cell divisions, guided by various molecular, cellular, and environmental signals at various developmental phases. In order to differentiate between symmetric and asymmetric cell division, an individual can observe dissimilarities in spindle orientation, or variable heritability of cytoplasmic or membrane-bound proteins including the cell fate determinant Numb and atypical protein kinase C [9 - 12]. Cells separate asymmetrically due to external or internal fate determinants: externally, different microenvironments will cause daughter cells to accept unlike fates; internally, cytoplasmic cell fate determinants (*e.g.*, Numb) would be asymmetrically confined in a cell and will segregate variably into daughters that accept different fates [13].

The mode of cell division: symmetric *versus* asymmetric cell divisions have the potential to sustain stem cell self-renewal to uphold the reservoir pool of ASCs; it is also crucial in order to keep an appropriate balance of self-renewal and differentiation of stem cells (Fig. **1**). For example, symmetric cell cytokinesis endows stem cells the ability to produce two daughter cells, both obtaining characteristics identical to the parent cell; such characteristics are necessary for developing stem cell pools if the daughter cells have stem cell fate. Furthermore, symmetric cell division has the ability to yield two daughters that attain a differentiation cell fate to produce differentiated somatic cells to maintain tissue function, and consequently contain less effectiveness than that of the mother stem cell (Fig. **1**). This phenomenon could result in quick construction of tissue "effector" cells, however there is also a chance it will lead to possible reduction of the stem cell reservoir [14, 15]. Asymmetric cell division (ACD), on the contrary, is vital in balancing self-renewal and differentiation. In addition, it is essential during development due to its ability to rectify spatial and temporal specification of cell lineages. In comparison to symmetric division, ACD yields two daughter cells that possess different cell fates. One daughter retains stem cell characteristics similar to the mother cell, and the other daughter formed by ACD loses stem cell characteristics, thereby becoming a differentiated somatic cell [16] (Fig. **1**). Symmetric *versus* asymmetric cell division will be discussed in more detail in sections 3 and 4.

Ku et al Figure 1

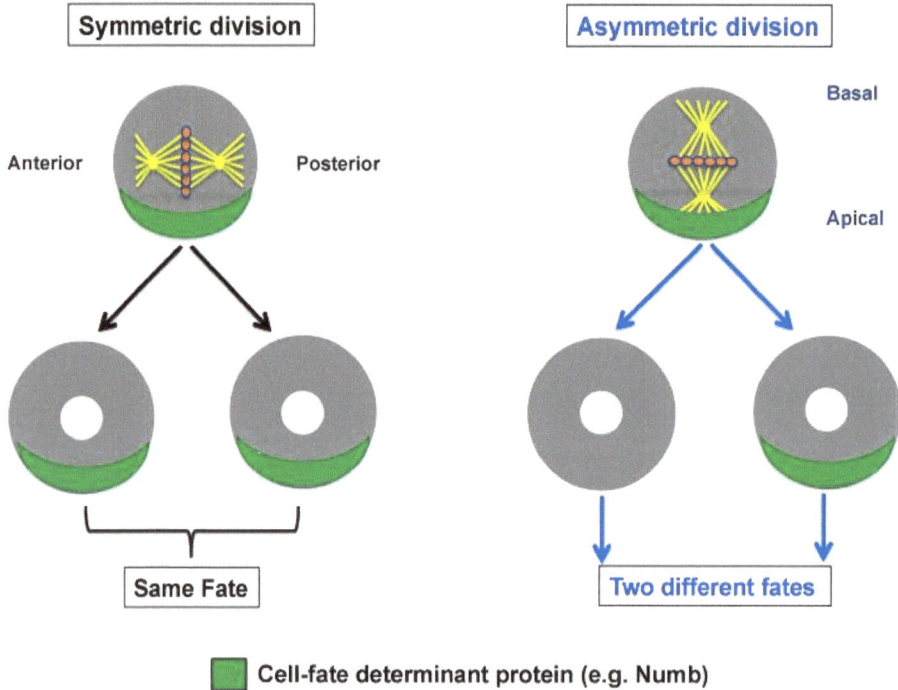

Fig. (1). Schematic depiction of a polarized dividing cell show two modes of cell division: symmetric and asymmetric cell division in epithelial cells. During symmetrical division, spindle orientation and determinant protein (*e.g.* Numb) localization are not coordinated. Determinants segregate equally, giving rise to two equal (stem) cells. During asymmetric division, spindle orientation and determinant protein (*e.g.* Numb) localization are coordinated, giving rise to a differentiating cell and a stem cell. Thus, the difference in Numb (green) expression levels between two daughter cells mediates asymmetric cell division, whereas lack of Numb inheritance by both daughters will allow them to execute the stem cell self-renewal program by maintaining Notch1 activity and thus allowing symmetric cell division, as we reported in distal lung epithelial stem cells [12]. [Adapted from [75]].

Another aspect of stem cell behavior is their life spans that depend on several intrinsic and extrinsic factors. Like accumulation of DNA damage, impairment of DNA repair mechanisms, oxidative stress caused by reactive oxygen-containing molecules, the shortening of telomeres, atypical deviations in gene expression propelled by age-related epigenetic shifts, and loss of function of the stem cell niche with age. Life spans of stem cells vary and depend on their types. For instance, while adult stem cells have a limited longevity because of their limited self-renewal and proliferative potential, both ES and cancer stem cells are able to undergo an infinite number of cell divisions, and maintain a stable telomere length as well as are relatively more resistant to DNA damage [17, 18].

3. SYMMETRIC *VERSUS* ASYMMETRIC CELL DIVISION

From bacteria to mammals, two types of cell division exist: symmetric and asymmetric. The former (*i.e.* symmetric division) holds the primary purpose of proliferation, thereby resulting in the expansion of cell populations. This process forms two equal daughter cells with identical developmental fates. On the other hand, asymmetric cell division (ACD) is a mechanism characteristic of stem cells. This process results in the formation of two daughter cells with unlike developmental fates: one daughter cell will separate along a specific lineage, while the other has the capacity to restore stem cell identity and resume dividing in an asymmetric manner. The cellular diversity present in all multicellular organisms is due to the cells' capacity for asymmetric division, which yields two different types of daughter cells.

The ACD provides the foundation for the construction of the body axis and cell fate determination during development. Furthermore, asymmetric divisions play a critical functional role in both the preservation of adult stem cells and production of a sufficient quantity of differentiating daughter cells. The latter acts to uphold tissue homeostasis and restoration. Maintenance of cell polarity is found to be crucial during asymmetric divisions because loss of cell polarity may lead to unregulated asymmetric divisions, which is frequently associated with enhanced stem cell self-renewal and tumorigenesis.

The asymmetric cell division (ACD) can be achieved by either intrinsic or extrinsic mechanisms. However, several recent studies have shown that the specification of distinct daughter cell fates during ACDs is regulated by a mixture of intrinsic and extrinsic processes. Asymmetric localization of cell-cell connections and/or intrinsic cell fate determinants and location in a particular environment ("niche") are instances of intrinsic mechanism employed to identify cell polarity and direct asymmetric divisions. During the process of mitosis, intrinsic mechanism incorporates the favored separation of cell fate determinants (*e.g.* Numb) into one of the two daughter cells. This indeed demands specific equipment that regulates correct spindle orientation and manages other important occurrences in this progression in order to ensure the successful segregation of determinants. Moreover, extrinsic mechanism includes cell–cell communication, and therefore the formation of dissimilar fates is fortified *via* signals from nearby cells. For instance, specification in regard to daughter cell fate depends on the relations between either daughter cells, or between a daughter cell and neighboring cells in metazoans. However, multiple genes regulate either processes of ACD directly or dictate the definitive cell fates of each daughter and ultimately regulate ACDs.

Asymmetry Stem Cell Divisions in Different Systems

In the animal kingdom, asymmetric stem cell divisions have been found in various tissue types. Yet, little is known about the elements and molecular processes that employ in order to individualize cell fate and position mitotic spindles during ACD. Interestingly, one of the typical processes of ACD employed in many stem cell systems is the initiation of the Notch signaling pathway and/or asymmetric segregation of the Notch inhibitor Numb appears.

Here, focus will be on examples of the asymmetric cell divisions dependent on Notch signaling engagement and appear in non-mammalian or mammalian systems.

Drosophila Intestinal Stem Cells

Drosophila intestinal stem cells (ISCs) are a well-investigated model organism for ACD. ISCs inhabit the intestinal basement membrane in masses of 2–3 basally situated diploid cells, which resemble a wedge on basement membrane, interspersed within polyploidy enterocytes (ECs). Several studies have shown that ISCs undergo self-renewal or proliferation to become enteroblast cells (EBs), quiescent progenitor cells that ultimately differentiate to ECs or enteroendocrine cells [19, 20].

A recent study by Ohlstein and Spradling [21] has revealed that, within the gut of adult flies, the Notch signaling mechanism is crucial in facilitating ISC asymmetric division. This study also has shown that the ISCs signal *via* Delta to trigger Notch target genes within the daughter enteroblasts. Therefore, despite Notch positivity of every cell in the clusters of stem cells, merely the IS in direct contact with the basement membrane will positively stain for the Notch ligand Delta, while Notch signaling is stimulated solely within the daughter enteroblast [21]. A relatively unknown topic deals with the molecular mechanisms resulting in occluding of Notch signaling activity in an ISC in order to facilitate this asymmetric division. Nonetheless, several studies have focused on the investigation of mitotic spindle orientation in ISCs that are dividing and confirmed that ISCs separate in a non-random fashion, such that one of the daughter ISCs remains an ISC if it has continued to be in association with the basement membrane, while the other daughter cell that was separated from the basement membrane undergoes differentiation to create an enteroblast [22]. How the mitotic spindle is orientated as well as the mechanism(s) regulating the process within ISCs have yet to be distinguished.

HEMATOPOIETIC STEM CELLS (HSCS)

Similar to Drosophila intestinal stem cells, numerous investigations propose the significance of Notch signaling in controlling the fate of HSCs by preventing differentiation. These studies demonstrate that transgenic mice carrying a Notch-responsive GFP reporter have the potential to be exploited in order to enhance for hematopoietic progenitors. Thus, using a transgenic Notch reporter strain, they show that GFP+ cells possessed roughly 40–60% HSCs. Furthermore, expression of GFP is considerably reduced within differentiating precursors [23, 24]. They further used the transgenic Notch reporter strain with real time imaging to picture hematopoietic precursor cell cytokinesis developing in culture. In addition, Duncan *et al*. (2005) and Wu *et al*. (2007) further utilized these investigation instruments in order to demonstrate that varying factors, including different kinds of oncogenic chromosomal translocations, BCR-ABL for instance, have the potential to influence the system of cell cytokinesis or cell production and the very existence of hematopoietic progenitors. These data clearly highlight the significance of extrinsic signals as well as intracellular influences in the regulation of hematopoietic precursor cell divisions. However, more work is necessary to elucidate the molecular processes incorporated in the control of ACD by such influences. Another interesting inquiry that remains uncovered is whether the niche and surrounding unlike cell types, such as osteoblasts, vascular endothelial cells and stromal reticular cells, effect the orientation arrangement of mitotic spindles and ACDs of HSCs as well. Moreover, despite an ability of hematopoietic progenitor cells to separate asymmetrically, a comparative significance of this type of division within blood homeostasis has not been established yet.

MUSCLE STEM CELLS

Satellite cells effectively act like muscle stem cells and dwell underneath the basement membrane, neighboring mature myofibers. These cells are usually dormant, however they have the potential to be stimulated to begin the cell cycle when injured. Satellite cells are necessary for sustaining the formation of myoblasts while postnatal development occurs as well as promoting muscle repair after injury. Several studies have reported the asymmetric separation of mature (immortal) and immature DNA strands to different daughter cells within dividing muscle-lineage cells during the occurrence of muscle development and restoration [25 - 27]. In addition, other *in vivo* and in culture studies indicated that stimulated expression of numerous differentiation genes as well as variance localization of cell fate proteins; such as Numb, in daughter cells. This provides support the proposition stating that muscle progenitors are able to endure ACDs [27 - 29]. The molecular mechanisms controlling muscle satellite cell division in the presence of diverse environmental signals, elements included in the process of specification of daughter cell fates, and the relative importance of ACD in muscle restoration has not been determined yet.

MAMMALIAN EPIDERMAL STEM/PROGENITOR CELLS

The epidermis exists as a stratified squamous epithelium creating a barricade to prevent entrance of detrimental microbes as well as preserve body fluids. In order to perform these essential roles, proliferative basal cells within the deepest stratum occasionally disconnect from the extracellular matrix basement membrane, migrate away and then ultimately die.

It is widely thought that many molecular aspects, which specify that ACDs are preserved through evolution. Numerous investigations dealing with mouse embryos *in vivo* as well as skin cells grown in culture have demonstrated support of symmetric and asymmetric cell division within epidermal progenitor cells in all mammals. Furthermore, Seery and Watt (2000) [30] have reported ACDs in the basal stratum in the esophageal epithelium. These studies have shown that progenitors found in the bottom of the epidermis have the ability to reproduce symmetrically in order to supply a greater amount of stem cells, and asymmetrically, to create stratified epithelium. In these epidermal progenitors, an ACD produces one multiplicative 'basal' cell that stays in association with the baso-lateral membrane as well as another disconnected 'supra-basal' cell, which lies shifted apically in the direction of the skin's exterior [31, 32]. Once supra-basal, cells halt the division process, and continue to transition into a differ-entiation process in order to create a barrier [33].

A leading study from Fuchs laboratory proposed that stratification of the skin happens *via* asymmetric cell divisions where the mitotic spindle positions perpendicularly to the basement membrane. This investigation has demonstrated that basal progenitors are tangibly associated with the underlying basement membrane; additionally, they possesspossess adhesion molecules, including integrins and cadherins, essential for spindle alignment. Connection of basal cells to the basement membrane likely results in a clustering of both integrins and growth factor receptors found at the bottom of cells that have the potential to effect stem cell behavior as well as maintenance. Lechler and Fuchs (2005) [31] further demonstrated that perpendicular cell divisions simulate a biological process to achieve imbalanced apportioning of signaling molecules that have their origin in the basement membrane into two daughter cells. Apical multiplexes of polarity proteins are, therefore, created away from the basement membrane and are then separated into a supra-basal cell, poised for differentiation within mammalian epithelial progenitors (Fig. **1**). Mitotic cells containing perpendicular spindles, which represent asymmetric divisions, possess an apical crescent of cortical LGN that is the mammalian Pins orthologue. LGN, in turn, connects with both mouse Inscuteable (mInsc) and Par3 at the apical cortex of each basal cells. Atypical PKCζ, aPKCζ, is another polarity protein that centralizes at the apical cortex area on the basal cells [31] (Fig. **1**). In addition, LGN binds the Mud orthologue NuMA, in turn tethering spindles located at the poles [34]. Lechler and

Fuchs (2005) have further shown that integrins and cadherins are vital for the occurrence of apical localization of aPKCζ, the Par3-LGN-Inscuteable complex and NuMA-dynactin to line up accordingly at the spindle within basal progenitor cells. However, the common theme is that the principal purpose of apical polarity proteins in mammalian epithelial cells is to establish mitotic spindle orientation and formation of apico-basal polarity, in contrast to the identification of stem cell fate [15, 31, 35 - 39].

Ku et al. Figure 2

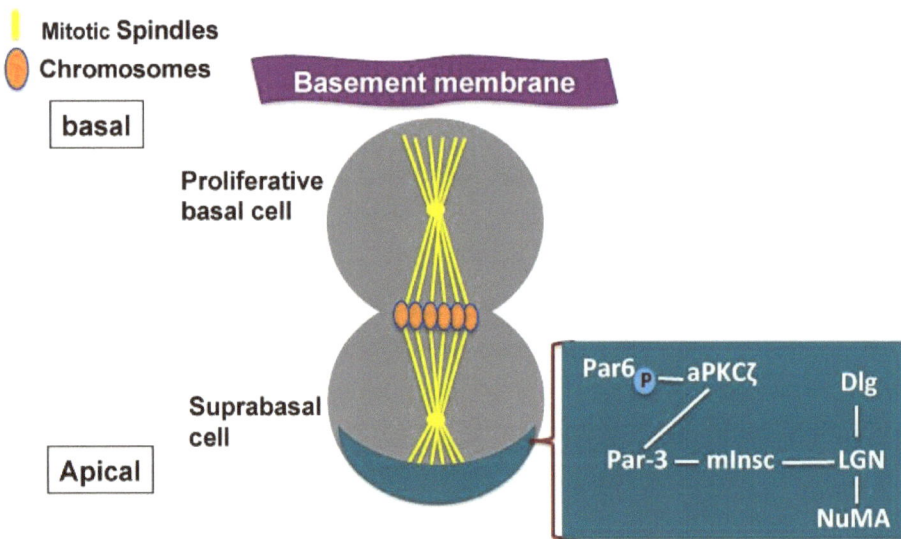

Fig. (2). Schematic depiction of a polarized mammalian mother cell during mitosis (anaphase) shows ACD in mammalian epithelia. Apical protein complexes are shown as a brown crescent. These apical protein complexes are important for both polarity establishment and spindle orientation in mammalian cells and are shown in a brown box, and described in the text. [Adapted from [75]].

Several other factors are also important of ACDs within mammalian epidermal progenitors. For instance, transcription factor p63 excites epidermal reproduction [40, 41]. In addition, it has been suggested that p63 a necessary factor for stratification because basal cells divide symmetrically only when p63 is not present [31, 42].

Other investigations propose the asymmetric triggering of Notch pathway as a process used in order to guarantee an asymmetric product in mammalian epithelial stem cell divisions, similar to Drosophila neuroblasts. Blanpain and colleagues [43] have demonstrated that supra-basal cells use Notch intracellular domain (NICD) with the goal of supporting differentiation. In addition, Smith *et al.* [44] have shown that the cell fate determinant Numb that also inhibits Notch signaling, centralizes predominantly to the baso-lateral cortex due to aPKC? Regulated phosphorylation, which leads to its elimination from the apical pole within cultured mammalian epithelial cells. Still, more investigations are needed to identify additional molecular processes as well as particular cell fate determinants meant to control ACDs, cell fate and behavior of mammalian epidermal stem cell while developing and mature stages.

MAMMALIAN NEURAL STEM/PROGENITOR CELLS

Several studies previously showed that symmetric as well as asymmetric cell divisions happen at varying developmental phases of neural, mammalian progenitors. In mammals, ACDs employ within the ventricular zone inside the cerebral cortex as well as neuroepithelium in the vertebrate retina [15, 45, 46]. Interestingly, symmetric cell divisions employ as well; they are largely noted during early developmental stages most likely in order to expand neural progenitor cell populations, unlike ACDs that occurred during later developmental stages with the goal of generating differentiating neurons.

Within vertebrate nervous systems, $G\alpha$-binding protein LGN, mouse Inscuteable (mInsc) as well as various additional influences have preserved functions in regulating spindle orientation like that of other systems. Additionally, Notch cues influence cell fate resolutions within the vertebrate nervous system [47 - 50]. Surprisingly, several studies suggest that Numb, which is a cell fate determinant and inhibits Notch signaling, could possibly effect cell fate *via* processes different than inhibiting Notch signaling activity [51, 52].

Numb, a cell fate determinant is an important factor that influences asymmetric *versus* symmetric cell division [16, 39]. Both ACD and Numb expression and purpose in development is a topic relatively well investigated in Drosophila and the nervous system of mammals [53], however, they are completely undistinguished within the lung.

Contrary to many systems in which spindle orientation acts as a gauge indicating if a given cell divides symmetrically or asymmetrically, the relationship between cell fat determination and mitotic spindle orientation is indistinctive in the maturation of vertebrate nervous system [54 - 57]. For instance, disturbance of the function of Gα-binding LGN, an essential mediator of mitotic spindle positioning, within the spinal cord neuroepithelium results in a random spindle positioning exclusive of disturbing daughter cell fate [56]. On the contrast, decrease of a different vital mediator of spindle positioning, called mouse Inscuteable (mInsc) and found within the retina results in disturbance of mitotic spindle positioning as well as a rise in the number of progenitor cells and neuronal deficiencies [55].

LUNG EPITHELIAL STEM/PROGENITOR CELLS IN MAMMALS

In mammals, regulation of epithelial stem and progenitor cells is critical in order to achieve adequate lung development [58, 59]. Fatal deficiencies of the capacity to diffuse gas, including the prevalent congenital types of lung hypoplasia and bronchopulmonary dysplasia (BPD), and the restricted ability of the lung to recuperate from such defects have the possibility to be clarified through a major deficit of stem and progenitor cells [59, 60]. Making sense of the way to attain an appropriate equilibrium between differentiation and self renewal/proliferation of lung-specific stem and progenitor cells is, therefore, essential because it is possible it will supply novel answers to reestablishing natural lung morphogenesis and potentially restoration of the gas diffusion surface. During development, asymmetric cell division is undoubtedly critical in maintaining an adequate balance between differentiation and self-renewal in addition to accurate temporal and spatial classification of cell lineages within the epithelium [15, 16].

Moreover, grasping the activity of lung epithelial stem and progenitor cells is of a major importance for identifying novel explanations for restoring usual lung morphogenesis. In addition, classification of ACD as well as determining innovative elements and processes controlling ACD as well as activity of lung epithelial stem/ progenitor cells, as primary processes that control the pendulum between progenitor cell self-renewal and differentiation within the lung, has the potential to aid in identifying new targets for the prevention and liberation therapy of lethal lung disease that occurs during infancy and childhood, as well as for lung restoration post-injury. Moreover, determination of the molecular mechanism controlling the relationship between the differentiation and proliferation of endogenous lung-specific progenitor cells is crucial in the development of methods aimed to make use of the capacity these cells contain to restore damaged and diseased lungs. Although this topic is imperative, there is still little known regarding ACD in epithelial stem/ progenitor cells within the lung.

Many previous articles propose that undifferentiated epithelial progenitors engage in numerous division-linked cell fate choices (symmetric and asymmetric) within the lung, leading to a seemingly homogeneous enlargement of the population of progenitor cells [61, 62]. At the time of embryonic development, multipotent epithelial stem and progenitor cells centralize in the outside of the lung epithelial buds/airways [63, 64]. For instance, recent investigations in this laboratory have shown that ACD probably regulates the equilibrium between lung epithelial stem/ progenitor cell conservation and differentiating populations of cells at outer epithelial ends. This laboratory has also provided novel support indicating that embryonic lung distal epithelial progenitors are polarized and extremely mitotic with distinctive perpendicular cell divisions. Perpendicular cell division is precisely related with ACD in varying mammalian epithelial cells due to the fact that they engage in asymmetric division *via* shifting of the spindle position from parallel to perpendicular [31]. Compatible with this newfound knowledge, mouse Inscuteable (mInsc), LGN (Gpsm2), and NuMA polarity proteins that control spindle positioning are asymmetrically centralized within mitotic distal epithelial progenitors of embryonic lungs [9]. Interfering with the purpose of the particular polarity proteins within lung epithelial cells *in vitro* unmethodically positions the spindle and alters cell fate [65].

In Drosophila as well as mammalian epithelial cells, ACD is regulated by the biased segregation of intrinsic cell fate determinants (CFDs) (*e.g.* Numb) to one of the daughter cells. CFDs are asymmetrically centralized when separating cells and dictate the axis of polarity responsible for deciding the position of the apical-basal cell division plane. This permits a quick shift from amplification, in which two like daughter cells are created, to diversification, in which daughter cells varying in shape are formed [53]. Numb protein, a Notch signaling inhibitor, is expressed consistently within the cytoplasm during interphase; however, it is centralized asymmetrically in splitting cells. Thus, Numb is distributed to just one of the two daughter cells, allowing this cell to accept an opposing fate compared to the other daughter cell. The cell accepting elevated amounts of Numb quells extrinsic Notch signaling and then differentiates, while the cell containing truncated Numb amounts preserves elevated Notch activity, thereby adopting stem cell fate [66 - 69]. In the embryonic lung, the cell fate determinant Numb, which is a vital determinant of asymmetric or symmetric cell division, is greatly expressed as well as asymmetrically dispersed at the apical side of outer epithelial progenitors [9, 12]. In addition, we also discovered that Numb is isolated to a single daughter cell in the majority of mitotic cells [9]. Therefore, the more perpendicular/asymmetric cell division is, the greater the likelihood that it will isolate Numb favorably to a single daughter cell in mitotic lung epithelial progenitors, strongly suggesting ACD within distal epithelial progenitors of embryonic lungs [12]. Decreasing Numb in MLE15 lung epithelial cells considerably raised the quantity of cells expressing the progenitor cell markers Sox9/Id2, aiding in its role as a cell fate determinant within the lung [12].

In several organs, epithelial cells distinctively demonstrate apical-basal polarity. In addition, they maintain a particular shape, in a way that a single and subtle alteration in cleavage plane from the standard position is enough to lead to an asymmetric instead of symmetric dispersal of the apical plasma membrane as well as adjacent adherent junctions to each daughter cell [8, 70]. When immunostaining for E-cadherin, a component of the apico-lateral junctional complex and lateral epithelial cell plasma membrane [71], the "cadherin hole" within the plasma membrane of mitotic epithelial cells seems like a fairly minute, unstained sector of the cell surface [12, 70]. Symmetric *versus* asymmetric dispersion of plasma membrane into each daughter cell has the potential to be predicted *via* the positioning of the cleavage plane in comparison to the cadherin hole within the epithelium of varying organs [70]. Recently, studies from our laboratory focused on cadherin hole studies of lung epithelium and showed that the majority of distal epithelial progenitors within embryonic lungs separate asymmetrically; using their cleavage, planes are estimated to detour the cadherin hole, leading to asymmetric dispersal of the cadherin hole into each daughter cell. Such discoveries supply and support for asymmetric cell division within distal epithelial progenitors of embryonic lungs [12].

Furthermore, studies in this laboratory have demonstrated that Eya1 protein phosphatase controls cell polarity, spindle positioning and centralization of Numb, a cell fate determinant that operates as an inhibitor of Notch signaling. Hence, Eya1 stimulates perpendicular division and Numb asymmetric isolation to only one daughter cell within mitotic distal lung epithelium, likely by controlling aPKC? phosphorylation concentrations. Therefore, epithelial cell polarity and mitotic spindle positioning are flawed after hindering Eya1 activity *in vitro* or *in vivo* [65]. Additionally, these studies have demonstrated that within Eya1$^{-/-}$ lungs, perpendicular separation is not preserved and Numb is dispersed to both daughters within mitotic epithelial cells, resulting in deactivation of Notch signaling. Furthermore, they demonstrated that genetic triggering of Notch signaling promoting progenitor cell character at the cost of differentiated cell phenotypes, has the potential to liberate the Eya1$^{-/-}$ lung phenotype, distinguished by loss of epithelial progenitors, increased epithelial differentiation with decreased branching.

Recent studies from our laboratory are, therefore, have shown that Eya1 protein phosphatase regulates the imperative balance between differentiation and self-renewal of distal lung epithelial progenitors by controlling ACD. This ACD-regulated balance is vital for longstanding preservation of tissue self-renewal at the time of development as well as in diseases. For example, congenital lung hypoplasia and bronchopulmonary dysplasia (BPD), in which a major deficit of stem/progenitor cells likely transpires, are prevalent characteristics of human prematurity and/or lung injury; thus, these are significant public health issues with human infancy. Adequate balance between differentiation and self-renewal of lung-specific progenitors, mediated by ACD, is unquestionably necessary for healthy lung morphogenesis as well as regeneration. Hence, regulated branching and outgrowth of epithelial tubes produces an adequately sizable gas diffusion surface to withstand life. Developmental deficiencies in this steady succession resulted in impaired differentiation as well as postnatal respiratory stress [58, 59].

In summary, recent studies from our laboratory give numerous supporting arguments that propose that asymmetric cell divisions are prevalent within embryonic distal lung epithelial progenitor cell populations. For instance, the cleavage plane positioning is estimated to detour the cadherin hole, leading to asymmetric dispersal of the cadherin hole to each daughter cell within the majority of outer epithelial progenitor cells [12]. Furthermore, this laboratory's article claims that the majority of distal epithelial cells contain apically centralized Par, LGN, NuMA, and mInsc polarity proteins, with mitotic spindles positioned perpendicular to the basement membrane as well as a distinguishing asymmetric segregation/inheritance of Numb [9], provides additional support that the cells divide asymmetrically. Certainly, a definitive relationship is present between ACD and the apical centralization of polarity proteins Par/LGN/NuMA/mInsc that regulates spindle positioning within epithelial mitotic cells in mammals [31], perpendicular placement of mitotic spindles, as well as asymmetric Numb dispersal within varying Drosophila and mammalian epithelial cell types [31], [72 - 74]. However, additional investigation are still required in order to resolve the question of asymmetric division in contrast to symmetric division within embryonic lung distal epithelial progenitors.

CONCLUDING REMARKS AND FUTURE DIRECTIONS

Recent studies that focused on asymmetric cell divisions through various organisms as well as within multiple stem cell systems have supplied ample understanding of different processes that are crucial to create cellular variety and preserve stem cells. Investigations utilizing invertebrate model schemes, including Drosophila, have found the significance of numerous extrinsic cues and intrinsic elements in stem cell division patterns; they have given standards for each of these cues and elements behave in order to distinguish asymmetric divisions. Several new investigations contribute evidence toward the notion that like processes are utilized within vertebrates. Yet, correct designation of stem cells *in vivo*, sophisticated seclusion of pure stem cell populations, and advances in real time imaging are still necessary in order to aid investigations contain the goal of recognition and characterization of the processes that regulate ACDs in more intricate mammalian stem cell systems, such as humans.

In addition, appropriate balance of the quantity of stem and progenitor cells is crucial in organ development, repair and regeneration. Several investigations addressing the processes controlling asymmetric stem cell divisions have emphasized the weight of balancing the number of stem cells. Proper balance and firm regulation of the quantity of stem and progenitor cells by asymmetric divisions are essential during the formation and preservation of tissues as well as during tissue repair and regeneration. This is due to a rise in the quantity of symmetric divisions that may be necessary for a finite length of time in order to increase the quantity of stem cells while tissue repair and regeneration is occurring. Importantly, numerous influences can hinder or possibly inhibit stem cells from returning from symmetric to asymmetric method of cell divisions. A good example of this is chronic injury or inflammation to a tissue that could damage the capacity for stem cells to react suitably to fix impaired tissues. It also is possible it will result in the inability of stem cells to shift from symmetric back to asymmetric method of divisions. Lack of success in adequate control of tissue reparation has the potential to ultimately result in the selection of stem cells resistant to natural growth control cues, a trademark of cancer cells. Consequently, grasping the processes that control asymmetric stem cell divisions increases the likelihood of designing effective tactics to block cancer initiation within varying cell types. This also could lead to identifying novel focuses for anti-cancer therapeutics. In addition, classification of the molecular processes and elements controlling the activity of adult stem cells is undoubtedly critical for the growth as well as the preservation of stem cells grown in culture, simultaneously preserving the stem cells' differentiation capacity. This could also lead toward directed differentiation of stem cells into varying specialized types of cells to be utilized in regenerative medicine.

CONFLICT OF INTEREST

The author confirms that author has no conflict of interest to declare for this publication.

ACKNOWLEDGEMENTS

The authors would like to acknowledge the support of research team at Children's Hospital Los Angles and Keck School of Medicine at the University of Southern California.

REFERENCES

[1] Ding S, Schultz PG. A role for chemistry in stem cell biology. Nat Biotechnol 2004; 22(7): 833-40.
[http://dx.doi.org/10.1038/nbt987] [PMID: 15229546]

[2] Daley GQ, Scadden DT. Prospects for stem cell-based therapy. Cell 2008; 132(4): 544-8.
[http://dx.doi.org/10.1016/j.cell.2008.02.009] [PMID: 18295571]

[3] Murry CE, Keller G. Differentiation of embryonic stem cells to clinically relevant populations: lessons from embryonic development. Cell 2008; 132(4): 661-80.
[http://dx.doi.org/10.1016/j.cell.2008.02.008] [PMID: 18295582]

[4] Takahashi K, Yamanaka S. Induction of pluripotent stem cells from mouse embryonic and adult fibroblast cultures by defined factors. Cell 2006; 126(4): 663-76.
[http://dx.doi.org/10.1016/j.cell.2006.07.024] [PMID: 16904174]

[5] Atala A, Lanza R, Thomson J, Nerem RM. Principles of Regenerative Medicine. Burlington, Mass, USA: Elsevier 2008.

[6] Hwang NS, Varghese S, Elisseeff J. Controlled differentiation of stem cells. Adv Drug Deliv Rev 2008; 60(2): 199-214.
[http://dx.doi.org/10.1016/j.addr.2007.08.036] [PMID: 18006108]

[7] Wodarz A. Establishing cell polarity in development. Nat Cell Biol 2002; 4(2): E39-44.
[http://dx.doi.org/10.1038/ncb0202-e39] [PMID: 11835058]

[8] Nelson WJ. Epithelial Cell Polarity From the Outside Looking In. News in physiological sciences: an international journal of physiology produced jointly by the International Union of Physiological Sciences and the American Physiological Society 2003; 18: 143-6.
[http://dx.doi.org/10.1152/nips.01435.2002]

[9] El-Hashash AH, Warburton D. Cell polarity and spindle orientation in the distal epithelium of embryonic lung Developmental dynamics : An official publication of the American Association of Anatomists 2011; 240(2): 441-5.
[http://dx.doi.org/10.1002/dvdy.22551]

[10] Huttner WB, Kosodo Y. Symmetric *versus* asymmetric cell division during neurogenesis in the developing vertebrate central nervous system. Curr Opin Cell Biol 2005; 17(6): 648-57.
[http://dx.doi.org/10.1016/j.ceb.2005.10.005] [PMID: 16243506]

[11] Morrison SJ, Kimble J. Asymmetric and symmetric stem-cell divisions in development and cancer. Nature 2006; 441(7097): 1068-74.
[http://dx.doi.org/10.1038/nature04956] [PMID: 16810241]

[12] El-Hashash AH, Warburton D. Numb expression and asymmetric *versus* symmetric cell division in distal embryonic lung epithelium. J histochem cytochem : official journal of the Histochemistry Society 2012; 60(9): 675-82.
[http://dx.doi.org/10.1369/0022155412451582] [PMID: 22713487]

[13] Yamashita YM. The centrosome and asymmetric cell division. Prion 2009; 3(2): 84-8.
[http://dx.doi.org/10.4161/pri.3.2.8821] [PMID: 19458491]

[14] Molofsky AV, Pardal R, Morrison SJ. Diverse mechanisms regulate stem cell self-renewal. Curr Opin

Cell Biol 2004; 16(6): 700-7.
[http://dx.doi.org/10.1016/j.ceb.2004.09.004] [PMID: 15530784]

[15]   Yamashita YM, Yuan H, Cheng J, Hunt AJ. Polarity in stem cell division: asymmetric stem cell division in tissue homeostasis. Cold Spring Harb Perspect Biol 2010; 2(1): a001313.
[http://dx.doi.org/10.1101/cshperspect.a001313] [PMID: 20182603]

[16]   Knoblich JA. Asymmetric cell division during animal development. Nat Rev Mol Cell Biol 2001; 2(1): 11-20.
[http://dx.doi.org/10.1038/35048085] [PMID: 11413461]

[17]   Reya T, Morrison SJ, Clarke MF, Weissman IL. Stem cells, cancer, and cancer stem cells. Nature 2001; 414(6859): 105-11.
[http://dx.doi.org/10.1038/35102167] [PMID: 11689955]

[18]   Rossi DJ, Jamieson CH, Weissman IL. Stems cells and the pathways to aging and cancer. Cell 2008; 132(4): 681-96.
[http://dx.doi.org/10.1016/j.cell.2008.01.036] [PMID: 18295583]

[19]   Micchelli CA, Perrimon N. Evidence that stem cells reside in the adult Drosophila midgut epithelium. Nature 2006; 439(7075): 475-9.
[http://dx.doi.org/10.1038/nature04371] [PMID: 16340959]

[20]   Ohlstein B, Spradling A. The adult Drosophila posterior midgut is maintained by pluripotent stem cells. Nature 2006; 439(7075): 470-4.
[http://dx.doi.org/10.1038/nature04333] [PMID: 16340960]

[21]   Ohlstein B, Spradling A. Multipotent Drosophila intestinal stem cells specify daughter cell fates by differential notch signaling. Science 2007; 315(5814): 988-92.
[http://dx.doi.org/10.1126/science.1136606] [PMID: 17303754]

[22]   Toledano H, Jones DL. Mechanisms regulating stem cell polarity and the specification of asymmetric divisions. 2009 Mar 31. StemBook. Cambridge, MA: Harvard Stem Cell Institute 2008. Internet

[23]   Duncan AW, Rattis FM, DiMascio LN, *et al.* Integration of Notch and Wnt signaling in hematopoietic stem cell maintenance. Nat Immunol 2005; 6(3): 314-22.
[http://dx.doi.org/10.1038/ni1164] [PMID: 15665828]

[24]   Wu M, Kwon HY, Rattis F, *et al.* Imaging hematopoietic precursor division in real time. Cell Stem Cell 2007; 1(5): 541-54.
[http://dx.doi.org/10.1016/j.stem.2007.08.009] [PMID: 18345353]

[25]   Cairns J. Mutation selection and the natural history of cancer. Nature 1975; 255(5505): 197-200.
[http://dx.doi.org/10.1038/255197a0] [PMID: 1143315]

[26]   Conboy MJ, Karasov AO, Rando TA. High incidence of non-random template strand segregation and asymmetric fate determination in dividing stem cells and their progeny. PLoS Biol 2007; 5(5): e102.
[http://dx.doi.org/10.1371/journal.pbio.0050102] [PMID: 17439301]

[27]   Shinin V, Gayraud-Morel B, Gomès D, Tajbakhsh S. Asymmetric division and cosegregation of template DNA strands in adult muscle satellite cells. Nat Cell Biol 2006; 8(7): 677-87.
[http://dx.doi.org/10.1038/ncb1425] [PMID: 16799552]

[28] Conboy IM, Rando TA. The regulation of Notch signaling controls satellite cell activation and cell fate determination in postnatal myogenesis. Dev Cell 2002; 3(3): 397-409.
[http://dx.doi.org/10.1016/S1534-5807(02)00254-X] [PMID: 12361602]

[29] Kuang S, Kuroda K, Le Grand F, Rudnicki MA. Asymmetric self-renewal and commitment of satellite stem cells in muscle. Cell 2007; 129(5): 999-1010.
[http://dx.doi.org/10.1016/j.cell.2007.03.044] [PMID: 17540178]

[30] Seery JP, Watt FM. Asymmetric stem-cell divisions define the architecture of human oesophageal epithelium. Curr Biol 2000; 10(22): 1447-50.
[http://dx.doi.org/10.1016/S0960-9822(00)00803-4] [PMID: 11102807]

[31] Lechler T, Fuchs E. Asymmetric cell divisions promote stratification and differentiation of mammalian skin. Nature 2005; 437(7056): 275-80.
[http://dx.doi.org/10.1038/nature03922] [PMID: 16094321]

[32] Smart IH. Variation in the plane of cell cleavage during the process of stratification in the mouse epidermis. Br J Dermatol 1970; 82(3): 276-82.
[http://dx.doi.org/10.1111/j.1365-2133.1970.tb12437.x] [PMID: 5441760]

[33] Fuchs E, Raghavan S. Getting under the skin of epidermal morphogenesis. Nat Rev Genet 2002; 3(3): 199-209.
[http://dx.doi.org/10.1038/nrg758] [PMID: 11972157]

[34] Du Q, Stukenberg PT, Macara IG. A mammalian Partner of inscuteable binds NuMA and regulates mitotic spindle organization. Nat Cell Biol 2001; 3(12): 1069-75.
[http://dx.doi.org/10.1038/ncb1201-1069] [PMID: 11781568]

[35] Macara IG. Par proteins: partners in polarization. Curr Biol: CB 2004; 14(4): R160-2.
[http://dx.doi.org/10.1016/j.cub.2004.01.048]

[36] Macara IG. Parsing the polarity code. Nat Rev Mol Cell Biol 2004; 5(3): 220-31.
[http://dx.doi.org/10.1038/nrm1332] [PMID: 14991002]

[37] Shin K, Wang Q, Margolis B. PATJ regulates directional migration of mammalian epithelial cells. EMBO Rep 2007; 8(2): 158-64.
[http://dx.doi.org/10.1038/sj.embor.7400890] [PMID: 17235357]

[38] Suzuki A, Ohno S. The PAR-aPKC system: lessons in polarity. J Cell Sci 2006; 119(Pt 6): 979-87.
[http://dx.doi.org/10.1242/jcs.02898] [PMID: 16525119]

[39] Knoblich JA. Asymmetric cell division: recent developments and their implications for tumour biology. Nat Rev Mol Cell Biol 2010; 11(12): 849-60.
[http://dx.doi.org/10.1038/nrm3010] [PMID: 21102610]

[40] Mills AA, Zheng B, Wang XJ, Vogel H, Roop DR, Bradley A. p63 is a p53 homologue required for limb and epidermal morphogenesis. Nature 1999; 398(6729): 708-13.
[http://dx.doi.org/10.1038/19531] [PMID: 10227293]

[41] Yang A, Schweitzer R, Sun D, *et al.* p63 is essential for regenerative proliferation in limb, craniofacial and epithelial development. Nature 1999; 398(6729): 714-8.
[http://dx.doi.org/10.1038/19539] [PMID: 10227294]

[42]   Senoo M, Pinto F, Crum CP, McKeon F. p63 Is essential for the proliferative potential of stem cells in stratified epithelia. Cell 2007; 129(3): 523-36.
       [http://dx.doi.org/10.1016/j.cell.2007.02.045] [PMID: 17482546]

[43]   Blanpain C, Lowry WE, Pasolli HA, Fuchs E. Canonical notch signaling functions as a commitment switch in the epidermal lineage. Genes Dev 2006; 20(21): 3022-35.
       [http://dx.doi.org/10.1101/gad.1477606] [PMID: 17079689]

[44]   Smith CA, Lau KM, Rahmani Z, *et al.* aPKC-mediated phosphorylation regulates asymmetric membrane localization of the cell fate determinant Numb. EMBO J 2007; 26(2): 468-80.
       [http://dx.doi.org/10.1038/sj.emboj.7601495] [PMID: 17203073]

[45]   Gönczy P. Mechanisms of asymmetric cell division: flies and worms pave the way. Nat Rev Mol Cell Biol 2008; 9(5): 355-66.
       [http://dx.doi.org/10.1038/nrm2388] [PMID: 18431399]

[46]   Neumüller RA, Knoblich JA. Dividing cellular asymmetry: asymmetric cell division and its implications for stem cells and cancer. Genes Dev 2009; 23(23): 2675-99.
       [http://dx.doi.org/10.1101/gad.1850809] [PMID: 19952104]

[47]   Chenn A, McConnell SK. Cleavage orientation and the asymmetric inheritance of Notch1 immunoreactivity in mammalian neurogenesis. Cell 1995; 82(4): 631-41.
       [http://dx.doi.org/10.1016/0092-8674(95)90035-7] [PMID: 7664342]

[48]   Petersen PH, Zou K, Krauss S, Zhong W. Continuing role for mouse Numb and Numbl in maintaining progenitor cells during cortical neurogenesis. Nat Neurosci 2004; 7(8): 803-11.
       [http://dx.doi.org/10.1038/nn1289] [PMID: 15273690]

[49]   Zhong W, Feder JN, Jiang MM, Jan LY, Jan YN. Asymmetric localization of a mammalian numb homolog during mouse cortical neurogenesis. Neuron 1996; 17(1): 43-53.
       [http://dx.doi.org/10.1016/S0896-6273(00)80279-2] [PMID: 8755477]

[50]   Zhong W, Jiang MM, Weinmaster G, Jan LY, Jan YN. Differential expression of mammalian Numb, Numblike and Notch1 suggests distinct roles during mouse cortical neurogenesis. Development 1997; 124(10): 1887-97.
       [PMID: 9169836]

[51]   Rasin MR, Gazula VR, Breunig JJ, *et al.* Numb and Numbl are required for maintenance of cadherin-based adhesion and polarity of neural progenitors. Nat Neurosci 2007; 10(7): 819-27.
       [http://dx.doi.org/10.1038/nn1924] [PMID: 17589506]

[52]   Zhou Y, Atkins JB, Rompani SB, *et al.* The mammalian Golgi regulates numb signaling in asymmetric cell division by releasing ACBD3 during mitosis. Cell 2007; 129(1): 163-78.
       [http://dx.doi.org/10.1016/j.cell.2007.02.037] [PMID: 17418793]

[53]   Betschinger J, Knoblich JA. Dare to be different: asymmetric cell division in Drosophila, *C. elegans* and vertebrates. Curr Biol 2004; 14(16): R674-85.
       [PMID: 15324689]

[54]   Sanada K, Tsai LH. G protein betagamma subunits and AGS3 control spindle orientation and asymmetric cell fate of cerebral cortical progenitors. Cell 2005; 122(1): 119-31.
       [http://dx.doi.org/10.1016/j.cell.2005.05.009] [PMID: 16009138]

[55]   Zigman M, Cayouette M, Charalambous C, *et al.* Mammalian inscuteable regulates spindle orientation and cell fate in the developing retina. Neuron 2005; 48(4): 539-45.
[http://dx.doi.org/10.1016/j.neuron.2005.09.030] [PMID: 16301171]

[56]   Morin X, Jaouen F, Durbec P. Control of planar divisions by the G-protein regulator LGN maintains progenitors in the chick neuroepithelium. Nat Neurosci 2007; 10(11): 1440-8.
[http://dx.doi.org/10.1038/nn1984] [PMID: 17934458]

[57]   Konno D, Shioi G, Shitamukai A, *et al.* Neuroepithelial progenitors undergo LGN-dependent planar divisions to maintain self-renewability during mammalian neurogenesis. Nat Cell Biol 2008; 10(1): 93-101.
[http://dx.doi.org/10.1038/ncb1673] [PMID: 18084280]

[58]   Warburton D, Perin L, Defilippo R, Bellusci S, Shi W, Driscoll B. Stem/progenitor cells in lung development, injury repair, and regeneration. Proc Am Thorac Soc 2008; 5(6): 703-6.
[http://dx.doi.org/10.1513/pats.200801-012AW] [PMID: 18684721]

[59]   Warburton D, El-Hashash A, Carraro G, *et al.* Lung organogenesis. Curr Top Dev Biol 2010; 90: 73-158.
[http://dx.doi.org/10.1016/S0070-2153(10)90003-3] [PMID: 20691848]

[60]   Shi W, Xu J, Warburton D. Development, repair and fibrosis: what is common and why it matters. Respirology 2009; 14(5): 656-65.
[http://dx.doi.org/10.1111/j.1440-1843.2009.01565.x] [PMID: 19659647]

[61]   Lu Y, Okubo T, Rawlins E, Hogan BL. Epithelial progenitor cells of the embryonic lung and the role of microRNAs in their proliferation. Proc Am Thorac Soc 2008; 5(3): 300-4.
[http://dx.doi.org/10.1513/pats.200710-162DR] [PMID: 18403323]

[62]   Rawlins EL. Lung epithelial progenitor cells: lessons from development. Proc Am Thorac Soc 2008; 5(6): 675-81.
[http://dx.doi.org/10.1513/pats.200801-006AW] [PMID: 18684716]

[63]   Rawlins EL, Hogan BL. Epithelial stem cells of the lung: privileged few or opportunities for many? Development 2006; 133(13): 2455-65.
[http://dx.doi.org/10.1242/dev.02407] [PMID: 16735479]

[64]   Rawlins EL, Clark CP, Xue Y, Hogan BL. The Id2+ distal tip lung epithelium contains individual multipotent embryonic progenitor cells. Development 2009; 136(22): 3741-5.
[http://dx.doi.org/10.1242/dev.037317] [PMID: 19855016]

[65]   El-Hashash AH, Turcatel G, Al Alam D, *et al.* Eya1 controls cell polarity, spindle orientation, cell fate and Notch signaling in distal embryonic lung epithelium. Development 2011; 138(7): 1395-407.
[http://dx.doi.org/10.1242/dev.058479] [PMID: 21385765]

[66]   Frise E, Knoblich JA, Younger-Shepherd S, Jan LY, Jan YN. The Drosophila Numb protein inhibits signaling of the Notch receptor during cell-cell interaction in sensory organ lineage. Proc Natl Acad Sci USA 1996; 93(21): 11925-32.
[http://dx.doi.org/10.1073/pnas.93.21.11925] [PMID: 8876239]

[67]   Guo M, Jan LY, Jan YN. Control of daughter cell fates during asymmetric division: interaction of Numb and Notch. Neuron 1996; 17(1): 27-41.

[http://dx.doi.org/10.1016/S0896-6273(00)80278-0] [PMID: 8755476]

[68]   Juven-Gershon T, Shifman O, Unger T, Elkeles A, Haupt Y, Oren M. The Mdm2 oncoprotein interacts with the cell fate regulator Numb. Mol Cellul Biol 1998; 18(7): 3974-82.
[http://dx.doi.org/10.1128/MCB.18.7.3974]

[69]   Yan B, Omar FM, Das K, Ng WH, Lim C, Shiuan K, *et al.* Characterization of Numb expression in astrocytomas. Neuropathol : Off J Jpn Soc Neuropathol 2008; 28(5): 479-84.
[http://dx.doi.org/10.1111/j.1440-1789.2008.00907.x]

[70]   Kosodo Y, Röper K, Haubensak W, Marzesco AM, Corbeil D, Huttner WB. Asymmetric distribution of the apical plasma membrane during neurogenic divisions of mammalian neuroepithelial cells. EMBO J 2004; 23(11): 2314-24.
[http://dx.doi.org/10.1038/sj.emboj.7600223] [PMID: 15141162]

[71]   Woods DF, Wu JW, Bryant PJ. Localization of proteins to the apico-lateral junctions of Drosophila epithelia. Dev Genet 1997; 20(2): 111-8.
[http://dx.doi.org/10.1002/(SICI)1520-6408(1997)20:2<111::AID-DVG4>3.0.CO;2-A]   [PMID: 9144922]

[72]   Cayouette M, Raff M. Asymmetric segregation of Numb: a mechanism for neural specification from Drosophila to mammals. Nat Neurosci 2002; 5(12): 1265-9.
[http://dx.doi.org/10.1038/nn1202-1265] [PMID: 12447381]

[73]   Haydar TF, Ang E Jr, Rakic P. Mitotic spindle rotation and mode of cell division in the developing telencephalon. Proc Natl Acad Sci USA 2003; 100(5): 2890-5.
[http://dx.doi.org/10.1073/pnas.0437969100] [PMID: 12589023]

[74]   Noctor SC, Martínez-Cerdeño V, Ivic L, Kriegstein AR. Cortical neurons arise in symmetric and asymmetric division zones and migrate through specific phases. Nat Neurosci 2004; 7(2): 136-44.
[http://dx.doi.org/10.1038/nn1172] [PMID: 14703572]

[75]   Berika M, Elgayyar ME, El-Hashash AH. Asymmetric cell division of stem cells in the lung and other systems. Front Cell Dev Biol 2014; 2: 33.
[http://dx.doi.org/10.3389/fcell.2014.00033] [PMID: 25364740]

CHAPTER 6

Adult Stem Cell Niches and Their Regulatory Molecular Mechanisms

John Ku[§], Wadah Alhassen[§], Haifen Huang[§], Salaheldin S. Soliman, Ahmed RN Ibrahim and **Ahmed HK El-Hashash[*]**

Stem Cells, Regenerative Medicine and Developmental Biology Program, Childrens Hospital Los Angeles, Keck School of Medicine and Ostrow School of Dentistry; University of Southern California, 4661 Sunset Boulevard, Los Angeles, CA 90027, USA

Abstract: The activation, survival, and quiescence of stem cells (SCs) are dependent on signaling within their niche or microenvironment. There are many types of SCs and SC niches that can be found in the human body. A single organ may contain more than one niche to accommodate for both slow and fast cycling SC populations. It appears that many SC niches possess similarities in both their cellular and molecular components. This chapter focuses on cellular organization, key molecular regulators, and the role of SC niches in aging and cancer.

Keywords: Autocrine, apoptosis, blood, canonical, extrinsic factors, growth factors, hair follicle, hematopoietic, intrinsic factors, intestine, lineage, microenvironment, mesenchymal, mechanism, neuron, paracrine, quiescence, regulation, stem cell niche, stromal, transduction.

INTRODUCTION

The idea of a stem cell niche or microenvironment has existed for over three decades [1]. Many studies have been performed showing that these SC niches regulate and maintain residential adult stem cells.

---

[*] **Correspondence author Ahmed El-Hashash:** Developmental Biology, Regenerative Medicine and Stem Cell Program, Saban Research Institute, Childrens Hospital Los Angeles, 4661 Sunset Boulevard MS 35, Los Angeles, California 90027, USA; Tel: 323-361-2764, 323-361-2258; Fax: 323-361-3613; E-mail: aelhashash@chla.usc.edu
[§] Equal first author

In mammalian systems, much work has been done with adult SCs in the hematopoietic system of the bone marrow, the hair follicles (HFs) of the skin, and the small intestine. High cell turnover rates in these three organs are dependent on the activity of fast-cycling SC populations [2 - 8]. Organs with slow turnover rates, such as the brain, muscle, and liver, contain only slow-cycling "reserve" SCs to maintain tissue and activate upon injury. Extrinsic signaling cues from the environment are important in maintaining SC quiescence and self-renewal, as well as survival of reserve SCs. These external inputs have been demonstrated by addition of such factors to culture medium for *in vitro* maintenance and activation of SCs [9]. Signals from more distant locations as well as from the SC itself or its progeny are also received, demonstrating the complexity of SC regulation.

1. ADULT SC NICHES

Cellular components of an SC niche is generally categorized into the following: (1) SCs and their progeny provide autocrine and paracrine regulation, respectively; (2) neighboring mesenchymal or stromal cells exhibit paracrine signals; (3) extracellular matrix (ECM) or cell-to-cell contact and; (4) external signals from blood vessels, neurons, and immune cells.

Regulatory Mechanisms of Intrinsic Stem Cells

Adult SC systems contribute to the niche through secreted signals and are directly influenced by the SC pool and its progeny [10]. Niches are crucial, as the behavior of hematopoietic stem cell (HSC) subpopulations is influenced by the bone marrow niche. SC progenies are also important as studies have demonstrated their role in maintaining HSC retention in the niche [11, 12].

Furthermore, a study investigating the role of lymphoid progeny revealed that fewer HSCs survived a transplant when regulatory T cells (Tregs) were depleted and supportive Treg stimuli were absent [13].

SCs and their progeny in hair follicles generate important niche signals. HFs frequently cycles through apoptosis and regeneration [14, 15]. Slow-cycling hair follicle stem cells (HFSCs) are located near the top of the follicle that remains intact during the apoptotic phase. During regeneration, fast-cycling SC progeny expand and differentiate until the next cycle. This cycle is thought to be dependent on exposure to quiescent and activating signals within the niche [3, 9, 16].

Intestinal stem cells (ISCs) are constantly being activated by signals from the niche as a result of high cell turnover rates. The two ISC populations are Lgr5+ (or CBC) and Bmi1+ (or +4 SCs); both have the ability to self-renew and give rise to all crypt cell lineages [2, 4]. Furthermore, studies have revealed that CBCs are capable of generating +4 SCs, depicting an importance in autocrine and niche regulation [5].

Satellite SCs are found to associate along the edge of muscle fibers [17]. Studies find that these satellite SCs are heterogeneous with fast and slow cycling SCs for tissue homeostasis and reserve SCs, respectively [18 - 20]. Little is known about the effect of niche signaling in these satellite SCs.

Two regions of the adult brain actively generate new cells. These regions include the subventricular zone (SVZ), which borders the lateral ventricles, and the subgranular zone (SGZ) located within the hippocampus [21]. Both regions contain glial fibrillary acidic protein-expressing radial glia-like cells that can generate neurons. The SVZ niche contains vascular cells, ependymal cells, astrocytes, and other differentiated SC progeny. Ependymal cells determine cell fate by producing gradients of morphogens through the beating of cilia [22]. The establishment of these niches have not been well understood despite studies that utilize neural stem cells (NSC) in the niche [23, 24].

Role of Neighboring Stromal Cells

In addition to stem cell progeny, surrounding connective tissue cells also influence the SC niche. These mesenchymal, or stromal, cells play a role in SC containment, regulation, and dispersion. In the bone marrow, for instance, nearby osteoblasts are responsible for maintaining a constant number of SCs in the niche. Experiments that have reduced or enhanced populations of osteoblasts have been met with decreases and increases in SC number, respectively [25 - 30]. This bone marrow niche is also influenced by nearby endothelial cells as well as CXCL12-abundant reticular (CAR) cells. CXCL12 is a chemokine that affect SC proliferation, differentiation, and retention in the bone marrow.

During the hair cycle, surrounding mesenchymal cells induce the posterior region of a HF, the dermal papilla (DP), to release signals necessary for successful regeneration of a new follicle. The exact mechanism by which this occurs is still unknown but studies have reinforced the necessity of the DP for regeneration [31]. In the intestinal SC niche, ISCs located at the base of intestinal glands are influenced by nearby pericryptic fibroblast cells and smooth muscle cells [32].

Functions of Extracellular Matrix and Physical Contact

SC niches are further influenced by components in the ECM and physical contact between cells in the niche. Physical contact between neighboring cells is just as essential as contact between these cells and the ECM. Adhesion molecules that facilitate these interactions impact SC dormancy, maintenance, and retention in the niche. These changes can be observed in several well-studied SC niche systems such as the HF, brain, and muscle. In the HF, adhesion molecules worth noting include integrin beta-1, linker protein alpha-catenin, and E-cadherin [33 - 35]. A lack of these adhesion molecules have been shown to detrimentally impact both the skin and the HF SC niche. In the brain, astrocytes, vascular cells, and E-cadherin directly influence the self-renewal and differentiation of neural SCs. Muscle SCs, on the other hand, are entirely enclosed in matrix. As a result, the ECM plays an important role in localizing various growth factors (GFs) that activate satellite cells, especially to sites of injury. These factors include hepatocyte GFs, fibroblast GFs, and intestinal GFs [36]. While cell-to-cell contact appears to be crucial to SC maintenance, studies have offered conflicting results as to which adhesion molecules influence which systems. SC niches are complex systems that are influenced by a variety of extracellular factors, factors that overlap and influence other systems. Signaling pathways in SC niches are difficult to investigate as a result of this complexity.

Functions of Distal Signals from the Macro-Environment

Components of the SC niche discussed so far involve SC progeny and neighboring stromal cells, but distant signals from within the same tissue or outside the tissue can also influence SC behavior in the niche. The HF SC niche, for instance, receive signals from the all-encompassing dermis of the body in order to regulate hair cycles in the skin [37]. In hematopoietic systems, neural input is comprehensive. Long distance signals can significantly impact the release of SCs from the bone marrow into the circulatory system. In the muscle, satellite cells are advantageously positioned near blood vessels where they have access to signals in systemic circulation [38, 39]. Interestingly enough, signals that reach the spermatogonial stem cell (SCC) niche from the circulatory system are prevented from leaking into other systems by Sertoli cell tight junctions. Systemic signals are thereby essential for the proper development of germ SCs.

2. MOLECULAR MECHANISMS CONTROLLING ADULT SC NICHES

Now that we have a basic understanding of how an adult SC niche is organized, we discuss molecular signals that affect SC niches. These regulators can be divided into two distinct categories, one composed of survival and maintenance signals and the other of activating signals.

Intrinsic Signals of Survival and Maintenance

Signals that regulate SC dormancy, survival, and self-renewal are largely extrinsic, but can be intrinsic as well. Intrinsic signals generally take advantage of cell cycling in order to establish and maintain SC quiescence. Halting cell cycles prevent HSC exhaustion [40]. Several proteins are involved in this process, a majority of which are tumor suppressors in the Rb protein family [41]. This protein family is quite redundant. As a result, a single deletion is not sufficient enough to induce proliferation. Several studies have had to abstract multiple proteins such as Rb, p130, and p107 before proliferation was observed [42 - 44].

In the HF SC niche, the Runx1/p21 complex is responsible for maintaining SC dormancy during the hair cycle [42]. In this complex, the effects of one protein lead to the inverse of the other. Runx1 represses the p21 tumor suppressor, which promotes the cell cycle and regeneration of a new follicle. If Runx1 is the one repressed, however, p21 is then upregulated and HF SC dormancy is maintained. Other factors crucial for SC survival in the hair follicle include Tbx1, NFATc1, and Lhx2 [45 - 47].

A variety of effector proteins have been identified in other SC niches such as the muscle, intestine, and central nervous system. These effector proteins either stimulate or inhibit SC growth and differentiation through signal transduction pathways such as Notch or Wnt.

Extrinsic Molecular Signals Regulating SC Quiescence and Self-Renewal

Different cell types act as extrinsic regulators of SCs within a niche. For example, TGF-β maintains inactivity of HSCs within the bone marrow niche with the help of glial cells [48, 49]. Oral administration of TGF-β1 also induces quiescence of ISC in the intestinal epithelium [50]. The TGF-β superfamily is known to contribute to SC quiescence through multiple signaling pathways.

The BMP family, part of the TGF-β superfamily, contribute to SC quiescence as well. Studies have shown that inactivation of BMPRIa causes HF SC activation and premature anagen [51]. BMP is also important in maintaining the integrity of the HF SC niche, evidenced by DP cells that lose the ability to initiate HF formation due to a deficiency of BMPRIa [52]. The resting phase of the hair cycle correlates with a high concentration of BMP ligands that promote dormancy of HF SCs [37]. The BMP pathway also prevents proliferation in the intestine; inactivation of BMPRIa in intestinal epithelium leads to expansion of the SC compartment by activating the canonical Wnt pathway [53, 54]. This same pathway is also responsible for NSC quiescence and niche size regulation for HSCs in the bone marrow [55, 56].

The canonical Wnt signaling pathway is important in SC homeostasis and shown to be necessary in maintaining HF SCs [57]. However, the role of Wnt signaling in the hematopoietic system is debated as Wnt ligands maintained HSC self-renewal *in vitro*, but not *in vivo* [58 - 61].

The Notch pathway is another important regulator of stem cells. It is thought to maintain HSCs in the bone marrow.Also, it controls cell proliferation in the intestines in a Wnt-dependent manner [62 - 64]. Notch has also been identified in maintaining NSCs in adults [65]. These three pathways are just some of many involved in regulating SC quiescence.

Activating Signals Required for Tissue Repair and Regeneration

The regeneration of tissue usually occurs through SC activation via external signals. TGF-β signaling not only maintains quiescence in organs, but is involved in SC activation as well. In the HF, TGF-β2, expressed by DP cells, allows for HF regeneration by repressing BMP signaling through activation of Smad2/3 [66].

The canonical Wnt pathway is important for HF SC proliferation and activation, previously demonstrated by β-catenin stabilization [67]. Studies have also attributed the role of Wnt signaling in myogenic differentiation and muscle regeneration [68]. Canonical Wnt signaling activates upon injury to regenerate muscle [69]. The Notch pathway, important in SC regulation and cell fate, promotes satellite cell proliferation during muscle regeneration, and is shown to be impaired in aged muscle [70].

Several secreted factors are linked to SC activation. Some examples include G-CSF, which activates HSCs and is also used to harvest cells from the bone marrow for transplants; Neuregulin1, which affects spermatogonia differentiation *in vitro* [71, 72]; endothelial cells expressing growth factors such as VEGF, which promotes satellite cell proliferation; and neurons secreting neurotrophins, NGF and BDNF, which are thought to influence satellite cell behavior [73].

3. DYSFUNCTION OF SC NICHE

Aging tissues show a lower capacity for repair. It appears that the gradual loss of SC function contributes to global tissue aging. In parabiotic studies, young and aged mice that shared blood circulation demonstrated that the serum in young mice revitalized aged SCs in older mice as it increased long-term HSC numbers in the bone marrow [74].

Another study investigated signaling changes in aged myofibers, showing that they emit increased levels of FGF2. This resulted in the production of additional FGF ligands in the niche, causing aging satellite SCs to produce fewer FGF pathway inhibitors, causing an increase in the proliferation and depletion of SC reserves [75]. The myofiber niche is therefore important in limiting the amount of soluble factors to maintain SC quiescence. The importance of the niche in aging is further illustrated in studies where diminishing germ SC function correlates with low fertility. Exposure SSCs to a developing niche environment restored activity, suggesting that germ SC and niche functionality declines with aging [76].

Cancer stem cells (CSCs) arise when normal SCs begin to proliferate uncontrollably. It is reasonable to assert that dysregulation within a niche could promote tumorigenesis. A study of stromal cells derived from human basal cell carcinomas found increased expression of Gremlin1, an antagonist of BMP; repression of BMP may allow for CSC expansion [77]. Dysregulation of the niche can influence the development of cancer, as shown in mice induced with myelodysplasia through targeting of niche cells for gene deletion in the bone marrow [78]. Studies of glioblastoma in particular, found that existing CSCs depend on interactions with nearby perivascular and immune niche cells to promote cancer cell survival and proliferation [79, 80]. CSCs may be capable of molding their own niche as they have in cutaneous squamous cell carcinomas; CSCs secrete signals that promote self-renewal and are capable of reshaping the vasculature of the niche [81].

CONCLUDING REMARKS

Stem cell niches are important in governing SC quiescence, maintenance, and activation. The three main signaling pathways identified in governing these behaviors include signals derived from TGF-β superfamily, the Wnt pathway, and the Notch pathway. Understanding the function of the SC niche is beneficial for future applications. A number of clinical trials have already utilized a niche-centric approach for bone marrow transplantation [82]. Better understanding of the SC niche will allow for better SC manipulation *in vitro* to create optimum systems for study. Furthermore, targeting the niche may be useful in both counteracting cancer-driven signals and activating tissue regeneration.

CONFLICT OF INTEREST

The author confirms that author has no conflict of interest to declare for this publication.

ACKNOWLEDGEMENTS

Special regards to David Warburton, MD, Children's Hospital Los Angeles.

REFERENCES

[1]   Schofield R. The relationship between the spleen colony-forming cell and the haemopoietic stem cell. Blood Cells 1978; 4(1-2): 7-25.
[PMID: 747780]

[2]   Barker N, van Es JH, Kuipers J, *et al.* Identification of stem cells in small intestine and colon by marker gene Lgr5. Nature 2007; 449(7165): 1003-7.
[http://dx.doi.org/10.1038/nature06196] [PMID: 17934449]

[3]   Greco V, Chen T, Rendl M, *et al.* A two-step mechanism for stem cell activation during hair regeneration. Cell Stem Cell 2009; 4(2): 155-69.
[http://dx.doi.org/10.1016/j.stem.2008.12.009] [PMID: 19200804]

[4]   Sangiorgi E, Capecchi MR. Bmi1 is expressed *in vivo* in intestinal stem cells. Nat Genet 2008; 40(7): 915-20.
[http://dx.doi.org/10.1038/ng.165] [PMID: 18536716]

[5]   Takeda N, Jain R, LeBoeuf MR, Wang Q, Lu MM, Epstein JA. Interconversion between intestinal stem cell populations in distinct niches. Science 2011; 334(6061): 1420-4.
[http://dx.doi.org/10.1126/science.1213214] [PMID: 22075725]

[6]   Takizawa H, Regoes RR, Boddupalli CS, Bonhoeffer S, Manz MG. Dynamic variation in cycling of hematopoietic stem cells in steady state and inflammation. J Exp Med 2011; 208(2): 273-84.
[http://dx.doi.org/10.1084/jem.20101643] [PMID: 21300914]

[7]   Wilson A, Laurenti E, Oser G, *et al.* Hematopoietic stem cells reversibly switch from dormancy to self-renewal during homeostasis and repair. Cell 2008; 135(6): 1118-29.
[http://dx.doi.org/10.1016/j.cell.2008.10.048] [PMID: 19062086]

[8]   Zhang YV, Cheong J, Ciapurin N, McDermitt DJ, Tumbar T. Distinct self-renewal and differentiation phases in the niche of infrequently dividing hair follicle stem cells. Cell Stem Cell 2009; 5(3): 267-78.
[http://dx.doi.org/10.1016/j.stem.2009.06.004] [PMID: 19664980]

[9]   Blanpain C, Lowry WE, Geoghegan A, Polak L, Fuchs E. Self-renewal, multipotency, and the existence of two cell populations within an epithelial stem cell niche. Cell 2004; 118(5): 635-48.
[http://dx.doi.org/10.1016/j.cell.2004.08.012] [PMID: 15339667]

[10]   Hsu YC, Fuchs E. A family business: stem cell progeny join the niche to regulate homeostasis. Nat Rev Mol Cell Biol 2012; 13(2): 103-14.
[http://dx.doi.org/10.1038/nrm3272] [PMID: 22266760]

[11]   Chow A, Lucas D, Hidalgo A, *et al.* Bone marrow CD169+ macrophages promote the retention of hematopoietic stem and progenitor cells in the mesenchymal stem cell niche. J Exp Med 2011; 208(2): 261-71.
[http://dx.doi.org/10.1084/jem.20101688] [PMID: 21282381]

[12]   Winkler IG, Sims NA, Pettit AR, *et al.* Bone marrow macrophages maintain hematopoietic stem cell

(HSC) niches and their depletion mobilizes HSCs. Blood 2010; 116(23): 4815-28.
[http://dx.doi.org/10.1182/blood-2009-11-253534] [PMID: 20713966]

[13] Fujisaki J, Wu J, Carlson AL, *et al. In vivo* imaging of Treg cells providing immune privilege to the haematopoietic stem-cell niche. Nature 2011; 474(7350): 216-9.
[http://dx.doi.org/10.1038/nature10160] [PMID: 21654805]

[14] Blanpain C, Fuchs E. Epidermal homeostasis: a balancing act of stem cells in the skin. Nat Rev Mol Cell Biol 2009; 10(3): 207-17.
[http://dx.doi.org/10.1038/nrm2636] [PMID: 19209183]

[15] Sennett R, Rendl M. Mesenchymal-epithelial interactions during hair follicle morphogenesis and cycling. Semin Cell Dev Biol 2012; 23(8): 917-27.
[http://dx.doi.org/10.1016/j.semcdb.2012.08.011] [PMID: 22960356]

[16] Tumbar T, Guasch G, Greco V, *et al.* Defining the epithelial stem cell niche in skin. Science 2004; 303(5656): 359-63.
[http://dx.doi.org/10.1126/science.1092436] [PMID: 14671312]

[17] Wagers AJ, Conboy IM. Cellular and molecular signatures of muscle regeneration: current concepts and controversies in adult myogenesis. Cell 2005; 122(5): 659-67.
[http://dx.doi.org/10.1016/j.cell.2005.08.021] [PMID: 16143100]

[18] Ono Y, Boldrin L, Knopp P, Morgan JE, Zammit PS. Muscle satellite cells are a functionally heterogeneous population in both somite-derived and branchiomeric muscles. Dev Biol 2010; 337(1): 29-41.
[http://dx.doi.org/10.1016/j.ydbio.2009.10.005] [PMID: 19835858]

[19] Ono Y, Masuda S, Nam HS, Benezra R, Miyagoe-Suzuki Y, Takeda S. Slow-dividing satellite cells retain long-term self-renewal ability in adult muscle. J Cell Sci 2012; 125(Pt 5): 1309-17.
[http://dx.doi.org/10.1242/jcs.096198] [PMID: 22349695]

[20] Relaix F, Zammit PS. Satellite cells are essential for skeletal muscle regeneration: the cell on the edge returns centre stage. Development 2012; 139(16): 2845-56.
[http://dx.doi.org/10.1242/dev.069088] [PMID: 22833472]

[21] Ming GL, Song H. Adult neurogenesis in the mammalian brain: significant answers and significant questions. Neuron 2011; 70(4): 687-702.
[http://dx.doi.org/10.1016/j.neuron.2011.05.001] [PMID: 21609825]

[22] Sawamoto K, Wichterle H, Gonzalez-Perez O, *et al.* New neurons follow the flow of cerebrospinal fluid in the adult brain. Science 2006; 311(5761): 629-32.
[http://dx.doi.org/10.1126/science.1119133] [PMID: 16410488]

[23] Carlén M, Meletis K, Göritz C, *et al.* Forebrain ependymal cells are Notch-dependent and generate neuroblasts and astrocytes after stroke. Nat Neurosci 2009; 12(3): 259-67.
[http://dx.doi.org/10.1038/nn.2268] [PMID: 19234458]

[24] Nomura T, Göritz C, Catchpole T, Henkemeyer M, Frisén J. EphB signaling controls lineage plasticity of adult neural stem cell niche cells. Cell Stem Cell 2010; 7(6): 730-43.
[http://dx.doi.org/10.1016/j.stem.2010.11.009] [PMID: 21112567]

[25] Calvi LM, Adams GB, Weibrecht KW, *et al.* Osteoblastic cells regulate the haematopoietic stem cell

niche. Nature 2003; 425(6960): 841-6.
[http://dx.doi.org/10.1038/nature02040] [PMID: 14574413]

[26]   Kiel MJ, Radice GL, Morrison SJ. Lack of evidence that hematopoietic stem cells depend on N-cadherin-mediated adhesion to osteoblasts for their maintenance. Cell Stem Cell 2007; 1(2): 204-17.
[http://dx.doi.org/10.1016/j.stem.2007.06.001] [PMID: 18371351]

[27]   Visnjic D, Kalajzic Z, Rowe DW, Katavic V, Lorenzo J, Aguila HL. Hematopoiesis is severely altered in mice with an induced osteoblast deficiency. Blood 2004; 103(9): 3258-64.
[http://dx.doi.org/10.1182/blood-2003-11-4011] [PMID: 14726388]

[28]   Wilson A, Trumpp A. Bone-marrow haematopoietic-stem-cell niches. Nat Rev Immunol 2006; 6(2): 93-106.
[http://dx.doi.org/10.1038/nri1779] [PMID: 16491134]

[29]   Zhang J, Niu C, Ye L, *et al.* Identification of the haematopoietic stem cell niche and control of the niche size. Nature 2003; 425(6960): 836-41.
[http://dx.doi.org/10.1038/nature02041] [PMID: 14574412]

[30]   Zhu J, Garrett R, Jung Y, *et al.* Osteoblasts support B-lymphocyte commitment and differentiation from hematopoietic stem cells. Blood 2007; 109(9): 3706-12.
[http://dx.doi.org/10.1182/blood-2006-08-041384] [PMID: 17227831]

[31]   Rompolas P, Deschene ER, Zito G, *et al.* Live imaging of stem cell and progeny behaviour in physiological hair-follicle regeneration. Nature 2012; 487(7408): 496-9.
[http://dx.doi.org/10.1038/nature11218] [PMID: 22763436]

[32]   Kosinski C, Li VS, Chan AS, *et al.* Gene expression patterns of human colon tops and basal crypts and BMP antagonists as intestinal stem cell niche factors. Proc Natl Acad Sci USA 2007; 104(39): 15418-23.
[http://dx.doi.org/10.1073/pnas.0707210104] [PMID: 17881565]

[33]   Raghavan S, Bauer C, Mundschau G, Li Q, Fuchs E. Conditional ablation of beta1 integrin in skin. Severe defects in epidermal proliferation, basement membrane formation, and hair follicle invagination. J Cell Biol 2000; 150(5): 1149-60.
[http://dx.doi.org/10.1083/jcb.150.5.1149] [PMID: 10974002]

[34]   Vasioukhin V, Bauer C, Degenstein L, Wise B, Fuchs E. Hyperproliferation and defects in epithelial polarity upon conditional ablation of alpha-catenin in skin. Cell 2001; 104(4): 605-17.
[http://dx.doi.org/10.1016/S0092-8674(01)00246-X] [PMID: 11239416]

[35]   Young P, Boussadia O, Halfter H, *et al.* E-cadherin controls adherens junctions in the epidermis and the renewal of hair follicles. EMBO J 2003; 22(21): 5723-33.
[http://dx.doi.org/10.1093/emboj/cdg560] [PMID: 14592971]

[36]   Yin H, Price F, Rudnicki MA. Satellite cells and the muscle stem cell niche. Physiol Rev 2013; 93(1): 23-67.
[http://dx.doi.org/10.1152/physrev.00043.2011] [PMID: 23303905]

[37]   Plikus MV, Baker RE, Chen CC, *et al.* Self-organizing and stochastic behaviors during the regeneration of hair stem cells. Science 2011; 332(6029): 586-9.
[http://dx.doi.org/10.1126/science.1201647] [PMID: 21527712]

[38]   Christov C, Chrétien F, Abou-Khalil R, *et al.* Muscle satellite cells and endothelial cells: close neighbors and privileged partners. Mol Biol Cell 2007; 18(4): 1397-409.
[http://dx.doi.org/10.1091/mbc.E06-08-0693] [PMID: 17287398]

[39]   Fukada S, Uezumi A, Ikemoto M, *et al.* Molecular signature of quiescent satellite cells in adult skeletal muscle. Stem Cells 2007; 25(10): 2448-59.
[http://dx.doi.org/10.1634/stemcells.2007-0019] [PMID: 17600112]

[40]   Pietras EM, Warr MR, Passegué E. Cell cycle regulation in hematopoietic stem cells. J Cell Biol 2011; 195(5): 709-20.
[http://dx.doi.org/10.1083/jcb.201102131] [PMID: 22123859]

[41]   Viatour P, Somervaille TC, Venkatasubrahmanyam S, *et al.* Hematopoietic stem cell quiescence is maintained by compound contributions of the retinoblastoma gene family. Cell Stem Cell 2008; 3(4): 416-28.
[http://dx.doi.org/10.1016/j.stem.2008.07.009] [PMID: 18940733]

[42]   Cobrinik D, Lee MH, Hannon G, *et al.* Shared role of the pRB-related p130 and p107 proteins in limb development. Genes Dev 1996; 10(13): 1633-44.
[http://dx.doi.org/10.1101/gad.10.13.1633] [PMID: 8682294]

[43]   LeCouter JE, Kablar B, Hardy WR, *et al.* Strain-dependent myeloid hyperplasia, growth deficiency, and accelerated cell cycle in mice lacking the Rb-related p107 gene. Mol Cell Biol 1998; 18(12): 7455-65.
[http://dx.doi.org/10.1128/MCB.18.12.7455] [PMID: 9819431]

[44]   Walkley CR, Orkin SH. Rb is dispensable for self-renewal and multilineage differentiation of adult hematopoietic stem cells. Proc Natl Acad Sci USA 2006; 103(24): 9057-62.
[http://dx.doi.org/10.1073/pnas.0603389103] [PMID: 16754850]

[45]   Chen T, Heller E, Beronja S, Oshimori N, Stokes N, Fuchs E. An RNA interference screen uncovers a new molecule in stem cell self-renewal and long-term regeneration. Nature 2012; 485(7396): 104-8.
[http://dx.doi.org/10.1038/nature10940] [PMID: 22495305]

[46]   Horsley V, Aliprantis AO, Polak L, Glimcher LH, Fuchs E. NFATc1 balances quiescence and proliferation of skin stem cells. Cell 2008; 132(2): 299-310.
[http://dx.doi.org/10.1016/j.cell.2007.11.047] [PMID: 18243104]

[47]   Rhee H, Polak L, Fuchs E. Lhx2 maintains stem cell character in hair follicles. Science 2006; 312(5782): 1946-9.
[http://dx.doi.org/10.1126/science.1128004] [PMID: 16809539]

[48]   Yamazaki S, Iwama A, Takayanagi S, Eto K, Ema H, Nakauchi H. TGF-beta as a candidate bone marrow niche signal to induce hematopoietic stem cell hibernation. Blood 2009; 113(6): 1250-6.
[http://dx.doi.org/10.1182/blood-2008-04-146480] [PMID: 18945958]

[49]   Yamazaki S, Ema H, Karlsson G, *et al.* Nonmyelinating Schwann cells maintain hematopoietic stem cell hibernation in the bone marrow niche. Cell 2011; 147(5): 1146-58.
[http://dx.doi.org/10.1016/j.cell.2011.09.053] [PMID: 22118468]

[50]   Puolakkainen PA, Ranchalis JE, Gombotz WR, Hoffman AS, Mumper RJ, Twardzik DR. Novel delivery system for inducing quiescence in intestinal stem cells in rats by transforming growth factor

beta 1. Gastroenterology 1994; 107(5): 1319-26.
[http://dx.doi.org/10.1016/0016-5085(94)90533-9] [PMID: 7926496]

[51]   Kobielak K, Stokes N, de la Cruz J, Polak L, Fuchs E. Loss of a quiescent niche but not follicle stem cells in the absence of bone morphogenetic protein signaling. Proc Natl Acad Sci USA 2007; 104(24): 10063-8.
[http://dx.doi.org/10.1073/pnas.0703004104] [PMID: 17553962]

[52]   Rendl M, Polak L, Fuchs E. BMP signaling in dermal papilla cells is required for their hair follicle-inductive properties. Genes Dev 2008; 22(4): 543-57.
[http://dx.doi.org/10.1101/gad.1614408] [PMID: 18281466]

[53]   Haramis AP, Begthel H, van den Born M, *et al.* De novo crypt formation and juvenile polyposis on BMP inhibition in mouse intestine. Science 2004; 303(5664): 1684-6.
[http://dx.doi.org/10.1126/science.1093587] [PMID: 15017003]

[54]   He XC, Zhang J, Tong WG, *et al.* BMP signaling inhibits intestinal stem cell self-renewal through suppression of Wnt-beta-catenin signaling. Nat Genet 2004; 36(10): 1117-21.
[http://dx.doi.org/10.1038/ng1430] [PMID: 15378062]

[55]   Mira H, Andreu Z, Suh H, *et al.* Signaling through BMPR-IA regulates quiescence and long-term activity of neural stem cells in the adult hippocampus. Cell Stem Cell 2010; 7(1): 78-89.
[http://dx.doi.org/10.1016/j.stem.2010.04.016] [PMID: 20621052]

[56]   Zhang J, Niu C, Ye L, *et al.* Identification of the haematopoietic stem cell niche and control of the niche size. Nature 2003; 425(6960): 836-41.
[http://dx.doi.org/10.1038/nature02041] [PMID: 14574412]

[57]   Huelsken J, Vogel R, Erdmann B, Cotsarelis G, Birchmeier W. beta-Catenin controls hair follicle morphogenesis and stem cell differentiation in the skin. Cell 2001; 105(4): 533-45.
[http://dx.doi.org/10.1016/S0092-8674(01)00336-1] [PMID: 11371349]

[58]   Reya T, Duncan AW, Ailles L, *et al.* A role for Wnt signalling in self-renewal of haematopoietic stem cells. Nature 2003; 423(6938): 409-14.
[http://dx.doi.org/10.1038/nature01593] [PMID: 12717450]

[59]   Willert K, Brown JD, Danenberg E, *et al.* Wnt proteins are lipid-modified and can act as stem cell growth factors. Nature 2003; 423(6938): 448-52.
[http://dx.doi.org/10.1038/nature01611] [PMID: 12717451]

[60]   Cobas M, Wilson A, Ernst B, *et al.* Beta-catenin is dispensable for hematopoiesis and lymphopoiesis. J Exp Med 2004; 199(2): 221-9.
[http://dx.doi.org/10.1084/jem.20031615] [PMID: 14718516]

[61]   Kirstetter P, Anderson K, Porse BT, Jacobsen SE, Nerlov C. Activation of the canonical Wnt pathway leads to loss of hematopoietic stem cell repopulation and multilineage differentiation block. Nat Immunol 2006; 7(10): 1048-56.
[http://dx.doi.org/10.1038/ni1381] [PMID: 16951689]

[62]   Kumano K, Chiba S, Kunisato A, *et al.* Notch1 but not Notch2 is essential for generating hematopoietic stem cells from endothelial cells. Immunity 2003; 18(5): 699-711.
[http://dx.doi.org/10.1016/S1074-7613(03)00117-1] [PMID: 12753746]

[63]   Fre S, Pallavi SK, Huyghe M, *et al.* Notch and Wnt signals cooperatively control cell proliferation and tumorigenesis in the intestine. Proc Natl Acad Sci USA 2009; 106(15): 6309-14.
[http://dx.doi.org/10.1073/pnas.0900427106] [PMID: 19251639]

[64]   Riccio O, van Gijn ME, Bezdek AC, *et al.* Loss of intestinal crypt progenitor cells owing to inactivation of both Notch1 and Notch2 is accompanied by derepression of CDK inhibitors p27Kip1 and p57Kip2. EMBO Rep 2008; 9(4): 377-83.
[http://dx.doi.org/10.1038/embor.2008.7] [PMID: 18274550]

[65]   Breunig JJ, Silbereis J, Vaccarino FM, Sestan N, Rakic P. Notch regulates cell fate and dendrite morphology of newborn neurons in the postnatal dentate gyrus. Proc Natl Acad Sci USA 2007; 104(51): 20558-63.
[http://dx.doi.org/10.1073/pnas.0710156104] [PMID: 18077357]

[66]   Oshimori N, Fuchs E. Paracrine TGF-β signaling counterbalances BMP-mediated repression in hair follicle stem cell activation. Cell Stem Cell 2012; 10(1): 63-75.
[http://dx.doi.org/10.1016/j.stem.2011.11.005] [PMID: 22226356]

[67]   Lowry WE, Blanpain C, Nowak JA, Guasch G, Lewis L, Fuchs E. Defining the impact of beta-catenin/Tcf transactivation on epithelial stem cells. Genes Dev 2005; 19(13): 1596-611.
[http://dx.doi.org/10.1101/gad.1324905] [PMID: 15961525]

[68]   Brack AS, Conboy IM, Conboy MJ, Shen J, Rando TA. A temporal switch from notch to Wnt signaling in muscle stem cells is necessary for normal adult myogenesis. Cell Stem Cell 2008; 2(1): 50-9.
[http://dx.doi.org/10.1016/j.stem.2007.10.006] [PMID: 18371421]

[69]   Otto A, Schmidt C, Luke G, *et al.* Canonical Wnt signalling induces satellite-cell proliferation during adult skeletal muscle regeneration. J Cell Sci 2008; 121(Pt 17): 2939-50.
[http://dx.doi.org/10.1242/jcs.026534] [PMID: 18697834]

[70]   Conboy IM, Conboy MJ, Smythe GM, Rando TA. Notch-mediated restoration of regenerative potential to aged muscle. Science 2003; 302(5650): 1575-7.
[http://dx.doi.org/10.1126/science.1087573] [PMID: 14645852]

[71]   Katayama Y, Battista M, Kao WM, *et al.* Signals from the sympathetic nervous system regulate hematopoietic stem cell egress from bone marrow. Cell 2006; 124(2): 407-21.
[http://dx.doi.org/10.1016/j.cell.2005.10.041] [PMID: 16439213]

[72]   Hamra FK, Chapman KM, Nguyen D, Garbers DL. Identification of neuregulin as a factor required for formation of aligned spermatogonia. J Biol Chem 2007; 282(1): 721-30.
[http://dx.doi.org/10.1074/jbc.M608398200] [PMID: 17098736]

[73]   Yin H, Price F, Rudnicki MA. Satellite cells and the muscle stem cell niche. Physiol Rev 2013; 93(1): 23-67.
[http://dx.doi.org/10.1152/physrev.00043.2011] [PMID: 23303905]

[74]   Conboy IM, Conboy MJ, Wagers AJ, Girma ER, Weissman IL, Rando TA. Rejuvenation of aged progenitor cells by exposure to a young systemic environment. Nature 2005; 433(7027): 760-4.
[http://dx.doi.org/10.1038/nature03260] [PMID: 15716955]

[75]   Chakkalakal JV, Jones KM, Basson MA, Brack AS. The aged niche disrupts muscle stem cell

quiescence. Nature 2012; 490(7420): 355-60.
[http://dx.doi.org/10.1038/nature11438] [PMID: 23023126]

[76]   Zhang X, Ebata KT, Robaire B, Nagano MC. Aging of male germ line stem cells in mice. Biol Reprod 2006; 74(1): 119-24.
[http://dx.doi.org/10.1095/biolreprod.105.045591] [PMID: 16177220]

[77]   Sneddon JB, Zhen HH, Montgomery K, *et al.* Bone morphogenetic protein antagonist gremlin 1 is widely expressed by cancer-associated stromal cells and can promote tumor cell proliferation. Proc Natl Acad Sci USA 2006; 103(40): 14842-7.
[http://dx.doi.org/10.1073/pnas.0606857103] [PMID: 17003113]

[78]   Raaijmakers MH, Mukherjee S, Guo S, *et al.* Bone progenitor dysfunction induces myelodysplasia and secondary leukaemia. Nature 2010; 464(7290): 852-7.
[http://dx.doi.org/10.1038/nature08851] [PMID: 20305640]

[79]   Charles N, Ozawa T, Squatrito M, *et al.* Perivascular nitric oxide activates notch signaling and promotes stem-like character in PDGF-induced glioma cells. Cell Stem Cell 2010; 6(2): 141-52.
[http://dx.doi.org/10.1016/j.stem.2010.01.001] [PMID: 20144787]

[80]   Filatova A, Acker T, Garvalov BK. Thecancerstemcellniche(s):Thecrosstalk between glioma stem cells and their microenvironment. Biochim Biophys Acta 2013; 1830: 2496-508.

[81]   Beck B, Driessens G, Goossens S, *et al.* A vascular niche and a VEGF-Nrp1 loop regulate the initiation and stemness of skin tumours. Nature 2011; 478(7369): 399-403.
[http://dx.doi.org/10.1038/nature10525] [PMID: 22012397]

[82]   Thomas A, Stein CK, Gentile TC, Shah CM. Isolated CNS relapse of CML after bone marrow transplantation. Leuk Res 2010; 34(4): e113-4.
[http://dx.doi.org/10.1016/j.leukres.2009.09.022] [PMID: 19811825]

CHAPTER 7

Stem Cell Regulatory Mechanisms During Wound Healing and Cancer

Ahmed RN Ibrahim[s], **Deshna Majmudar**[s], **Safia Gilani**[s], **Jesse Garcia Castillo** and **Ahmed El-Hashash**[*]

Stem Cells, Regenerative Medicine and Developmental Biology Program, Children's Hospital Los Angeles, Keck School of Medicine and Ostrow School of Dentistry, University of Southern California, 4661 Sunset Boulevard, Los Angeles, CA 90027, USA

Abstract: Stem cells are known to have a paramount function in tissue regeneration and also the proliferation of cancer, and the ability that stem cells have to self renew allows them to differentiate and to regenerate injured tissues. More importantly, this capacity to self renew allows cancer stem cells to proliferate and promote cancer. Mesenchymal stem cells are multipotent cells that have the ability to differentiate into various cell types including; adipocytes, osteoblasts, and chondrocytes. These cells are known to regulate the healing of injuries and wounds and to activate cancer growth by secreting bioactive factors through paracrine signaling. Through scientific research, there is evidence that tissue specific and cancer stem cells also affect their surroundings through paracrine mechanisms, which would permit stem cells to facilitate wound recuperation and tumor proliferation, respectively. Because of this important connection, further investigation of the paracrine mechanisms by stem cells would ameliorate cancer treatment and cast light on the mechanisms of tissue regeneration.

Keywords: Cancer, Stem Cells, Tissue Injury, Wound Healing.

INTRODUCTION

Stem cells are undifferentiated biological cells that undergo self- repair and also

[*] **Correspondence author Ahmed El-Hashash:** Developmental Biology, Regenerative Medicine and Stem Cell Program, Saban Research Institute, Children's Hospital Los Angeles, 4661 Sunset Boulevard MS 35, Los Angeles, California 90027, USA; Tel: 323-361-2764, 323-361-2258; Fax: 323-361-3613; E-mail: aelhashash@chla.usc.edu
[s] First equal author

differentiate into many different cell types [1], in addition to having the ability to heal wounds and maintain tissue homeostasis [2]. They are also involved in wound healing [3]. Recently, it has been shown that the presence of higher number of stem cells in the injured area of hair follicle may accelerate the wound healing [4]. There is a resounding notion that the differentiation capacity of stem cells is paramount in wound recuperation, but other evidence has shown that paracrine signaling of stem cells plays a role in wound recuperation [5].

One of such stem cells that is known to have paracrine mechanisms in wound repair is the mesenchymal stem cell (MSC) that originates primarily in the adipose tissue or bone marrow [6, 7]. MSCs are activated by wounds and injuries and then proceed to regulate inflammatory responses and activate tissue repair. Initially, MSCs were known to simply differentiate into cells necessary for tissue repair, such as epithelial cells or nerve cells [8 - 11]. Subsequent studies proved that the paracrine functions of MSCs were more significant in tissue repair than differentiation potential [12 - 14].

Paracrine signaling allows the MSCs to induce changes in nearby cells, therefore altering the behavior or differentiation of the affected cells. Cancer stem cells (CSCs) are thought to function in tumor initiation and metastasis, in addition to their capability to attract MSCs [15 - 17]. CSCs allow tumor heterogeneity through the proliferation of multiple forms of differentiated cells. For example, breast cancer CSCs generate the basal as well as luminal breast cancer cells [18]. As is said for MSCs, CSCs may also act through paracrine mechanisms that impact cancer expansion and maintenance besides recruitment of differentiated cells. Since there are already many reviews on the paracrine signaling role of MSCs in tissue repair and cancer proliferation [12, 19 - 24], more focus will be placed on the paracrine functions of non-MSC stem cells. MSC paracrine function will be used only for comparisons.

As mentioned above, cells can communicate through paracrine signaling over a short distance by inducing changes in nearby cells through the release of paracrine factors. One way in which cells communicate *via* paracrine factors is through proteins such as cytokines and growth factors [22, 23, 25]. MSCs secrete a large amount of these proteins; some are very important for the survival of the nearby differentiated cells whereas others are stimulatory factors for angiogenesis and the development of new blood cells [26]. This amalgamation of proteins released by the cells is known as the secretome and other non-protein molecules such as lipids and RNAs can be secreted into the extracellular space. Many of these additional non-protein molecules, such as RNAs leave in packets of micro-vesicles generated by the cell that they are secreted from. Micro-vesicles, which are circular fragments generated from endosomes or plasma membranes are different from apoptotic bodies because of their obvious lack of both DNA and histones [27 - 29].

When nearby cells take up the micro-vesicles derived from MSCs, they can take the micro-vesicular contents and use them for various biological activities [30, 31]. Micro-vesicular RNA specifically is of importance. This RNA can be taken up by cells and translated into proteins which may affect the overall function of the recipient cell [32]. Therefore, micro-vesicles are effective in that they prevent apoptosis, activate stem cell activity and regulate inflammatory responses, which show how important this type of paracrine signaling may be on cell regeneration and differentiation [33 - 35].

STEM CELLS AND TISSUE INJURIES

Stem Cells and Cardiac Injuries

Cardiac stem cells have been demonstrated to ameliorate the myocardia after myocardial ischemia by differentiating and replacing injured cells. Recently, a study has shown that the differentiating ability for cardiac stem cells is not sufficient to help recovery and repair [36]. Additional paracrine signal molecules are necessary to activate pathways that enable cardiac stem cells to aid in the recovery of the heart after ischemia. For example, the secretory molecule named signal transducer and activator of transcription 3 (STAT3) must be stimulated in order for the cardiac stem cells to exhibit their cardio-protective effect. However, the activation of STAT3 occurs when the chemokine called stromal cell derived factor-1 (SDF-1) is released by the individual cardiac stem cells [37]. To exhibit the importance of SDF-1 in recovery of the myocardial infarction, SDF-1 was inhibited. When SDF-1 was inhibited, recovery was blocked. SDF-1 has two roles during myocardial repair. SDF-1 acts as a signal that directs stem cells to the infarcted heart [37] and also ameliorates the survival rates of cardiomyocytes [38] by suppressing caspase 3-dependent apoptosis. In the canine myocardial infarction model, recruitment of cardiac stem cells to the infarcted heart was achieved by administration of hepatocyte growth factor (HGF) and insulin growth factor-1 (IGF-1). These two growth factors have been previously demonstrated to help in the proliferation of cardiac stem cells [39]. Additionally, mesenchymal stem cells (MSCs) are also able to ameliorate the myocardial post-ischemia [40]. It was originally thought that the multipotent MSCs can be differentiated into cardiomyocyte-like cells which drove the recovery of myocard, however, it was found out that presence of a specific cocktail of secretory proteins was actually sufficient to drive recovery of the myocard [5, 41, 42]. Similarly to cardiac stem cells, MSCs also stimulate phosphorylation of STAT3 when present in the myocard [43]. When toll-like receptor 4 (TCL 4) was deficient in MSCs, there were higher levels of STAT3 activation, which increased the repairing rate of myocardial tissue compared to the wild type MSCs. When the MSCs were incubated in MSC- conditioned media (CM), the levels if SDF-1 in the infracted heart were higher [44]. The SDF-1 levels could have elevated when the CM was collected from MSCs culture that had been stimulated for vascular endothelial growth factor (VGEF) expression. Because some of the SDF-1 protein was produced by the MSCs and some was produced by the myocard, both MSCs and cardiac stem cells may express their cardioprotective property through the same pathways induced by these secretory proteins. In the porcine model, human embryonic derived MSCs also demonstrated improved recovery post-ischemia depending on the molecules these cells secreted [45]. This study also demonstrated that there was a decrease in the phosphorylation of Smad2 when the cardioprotective effect was exhibited by the MSCs as well as declined expression levels of caspase 3. The component that was responsible for the influence of the MSC-derived CM was a large complex of >1,000 D. This unknown component was later identified to be a 20S proteasome that mediated MSC-dependent cardioprotection [46]. The uptake of the proteasome by cardiomyocytes declined the concentration of mis-folded proteins and as a result, the survival rate of these cells increased. This agrees with the notice that MSC-derived CM increased the level of Bcl2, an anti-apoptotic protein thus ameliorates hypoxia-induced apoptosis in cardiomyocytes [47]. MSCs can also activate angiogenesis in the infracted myocard. MSC-derived CM was found to stimulate endothelial cells and also increased capillary density of the infracted heart [42, 48]. The blocking of VGEF and basic fibroblast growth factor (bFGF) by antibody treatment demonstrated that these two factors were vital pro-angiogenic factors [42, 47, 49, 50] since blocking these two factors caused low levels of recovery by MSCs [50]. Another pro-angiogenic factor, cysteine-rich angiogenic inducer 61 (Cyr61) has been found to have a role in the stimulation of angiogenesis of MSCs in the infarcted heart [51]. MSCs anti-fibrotic activity has also been found to participate to be useful the infarcted heart. MSC- derived CM decreased cardiac fibrosis by hindering cardiac fibroblast proliferation, which also caused the decreased deposition of collagen I, II, and III [52, 53]. There are two additional stem/progenitor cell lines that exhibit the same pathways activated by the same secretory molecules that help in the recovery of an infarcted myocard. These two cell lines are the endothelial progenitor cells, derived from the bone marrow, and the skeletal muscle derived stem cells (MDSC), which demonstrated the cardioprotective effect that the MSCs and the cardiac stem cell exhibited [54, 55]. The bone marrow-derived endothelial progenitor cell EPCs transplanted into the myocard also exhibited increased expression of the chemokine, SDF-1 [54]. EPCS in the myocard can also undergo angiogenesis by expressing thymosin B4 that has been shown to improve endothelial function [56]. MDSC, however, had low levels of differentiation into cardiomyocytes when they were placed in an infarcted heart [57]. Although they were not able to differentiate as efficiently, their secretory activity helped the recovery process. The major secretory molecule that MDSCs secreted is VEGF, which stimulates angiogenesis. Blocking VGEF resulted in decreased neovascularization and unfavorable remodeling. The stretching of these MDSCs induced the secretion of VEGF. Additionally, the murine model demonstrated that exercising after infarction increased VGEF levels and angiogenesis [58]. This demonstrates the physical therapy post-ischemia enables the activation of VGEF-dependent neovascularization and improves recovery [59]. In a study comparing the cardioprotective effects of both MSCs and EPCs, cardiosphere-derived cells were more efficient in enhancing both differentiation of myocytes and tube formation. These cardiosphere-derived cells are a type of cell population that contains cardiac stem cells as well as supporting cells [60]. These cells had higher levels of secretory molecules such as SDF-1, HGF, VGEF, and bFGF compared to MSCs and bone marrow-derived mononuclear progenitor cells.

Duren *et al.* investigated whether or not the protein cocktail would change when both MSCs and cardiac stem cells were transplanted into the myocard at the same time [61]. They started this study by culturing the cells *in vitro*, where both cells had managed to secrete all the eight molecules they had measured [61]. SDF-1 and VGEF were among the eight secretory factors that they tested. However, *in vivo* these cells did not secrete all 8 of the secretory factors they had initially tested. In fact, the two dominant secretory molecules were VGEF and bFGF. One significant different from *in vivo* and *in vitro* were that these cells never secreted SDF-1 after transplantation. Both cell lines also stimulated neovascularization in the infracted area. Cardioprotective stem cells secrete multiple factors (chemokines, growth factors, pro-angiogenetic factors, *etc.*) that ameliorate the survival of cardiomyocytes, and induce neovascularization.

Stem Cells and Nervous System Damages

Recovery of an ischemic brain also needs multiple secreted factors of stem cells. The pro-angiogenic factor, VGEF, again plays a role in the development of stem cell–dependent repair damaged caused by stroke- induced lesions [62]. Neural stem cells also stimulate axonal transport and enhanced dendritic branching and length [63]. The secretion of bospondin 1 and 2 caused neuronal stem cell-induced plasticity of dendrites by the transplanted stem cells. Bospondin 1 and 2 roles in stroke recovery were determined by analyzing bospondin 1 and 2 knockout mice and their decreased stroke recovery rates [64]. In murine models, the neural stem cells also improved the repair of spinal cord injuries. Furthermore, when these cells were transplanted into the lesion areas, these cells improved axonal growth [65]. Enhanced axonal growth is caused by the secretion of neutrotrophic factors nerve growth factors (NGF) and brain derived neutrotrophic factors (BDNF) by the stem cells. When CM taken from bone marrow derived MSCs was injected into rats with spinal cord injuries, the rats had improved motor recovery [66]. Even though MSCs secrete NGF and BDNF, protect neurons from apoptosis [66] and induce neurite overgrowth [67], MSC-CM does not have an effect on axonal outgrowth *in vivo*. MSC-CM appears to exhibit its neuroprotective character *in vivo* by enhancing angiogenesis, which is dependent on VGEF.

Stem Cells and Kidney Injury/Recovery

Paracrine secretory molecules also play a crucial role in stem cell mediated kidney recovery. Tubular adult renal stem/ progenitor cells (tARPC) promote proliferation and suppress the apoptosis of cisplatin-induced toxicity in proximal tubular epithelial cells [68]. Proliferation and inhibition of apoptosis depended on the secretion of inhibitin A, which is known to regulate renal tubulogenesis by inhibiting the process [69]. The inhibin A RNA is moved to tubular epithelial cells *via* micro-vesicles [68]. However, inhibin A is just found in micro-vesicles that originated from tARPC that had come into contact with damaged tubular epithelium. Toll receptor 2 (TLR2) is a necessary receptor in the recognition of apoptotic tubular epithelial cells by tARPC. Besides tARPC, micro-vesicles that were synthesized by bone marrow derived mesenchymal cells increased survival and proliferation of damaged epithelial cells.

MSC-derived micro-vesicles and tubular cells communicate by the presence of CD44 and CD29 receptors on the outer surface of these micro-vesicles. The renoprotective effect of MSCs may have been caused by the secretion of paracrine factors such as VGEF, IGF-1, and HGF by MSCs.

These soluble factors maybe the responsible for the improved survival of endothelial cells when MSCs are present [70, 71]. MSCs interact to endothelial cells to form tubes in a cooperative way [70]. In chronic kidney disease, exosomes derived from MSCs did not demonstrate any improvement of the diseases [72]. However, non-fractioned MSC-derived CM reduced the rate of progression of the disease and conserved renal function.

Stem Cells and Other Tissue Injuries

Stem cell-derived secretory factors also play a crucial role in the recovery of the liver from cirrhosis [73]. To determine whether or not the secretory factors in CM played a role in liver recovery, dimethylnitrosamine-induced liver injury in rats was treated with CD34+ hematopoietic stem cell-derived CM. After the injection of the CM into the tail vein of rats, there was a significant increase in liver recovery and subsequently animal survival. The blockade of caspase-3-dependent apoptosis of hepatic cells caused the increased survival rate and liver recovery. Cytokines belonging to the CXCL family, which is known to participate in wound repair, was one of 32 factors found in the CM of CD34+ hematopoietic stem cells. Liver repair is closely linked to CXC receptor 2 [74], which recognizes CXCL cytokines.

PARACRINE EFFECT OF STEM CELLS IN CANCER

Glioma

CD 133+ glioma cancer-initiating/stem-like cells (CSCs) are able to inhibit immune responses against the tumor. This effect is produced by inhibiting T-cell effector activity [75]. These activities depend on presences of Phosphorylated STAT3 in the CSCs, and capability to stimulate STAT3 in the immune cells. Since CM from the glioma CSCs was as effective as the CSCs themselves in inducing immunosuppression [76], it is likely that CSC-secreted molecules are responsible for STAT3 stimulation. Among the molecules, transforming growth factor beta 1 and prostaglandin E2, are the two major secretory molecules responsible for the immunosuppression of MSCs [24]. Additionally, galectin-3, a beta galactosides-binding protein which in its soluble form can activate T-cell apoptosis [77].

This protein is secreted by glioma CSCs [76]. Interestingly, galactin -3 is produced by glioma cells, but not by oligodendrocytes or astrocytes [78]. The glioma CSCs also produces the B7-H1 protein on its surface [76]. This protein suppressed T-cell proliferation through cell-cell interaction. These data demonstrated that glioma CSCs have a significant contribution to the immunosuppression in glioblastoma multiforme. This immunosuppression is caused by paracrine effects of glioma CSCs as well as by mechanisms including direct contacts with immune cells.

CSCs were also found to promote angiogenesis: CSCs secrete a large amount of VEGF, which enhances endothelial migration as well as tube formation [79].

Hypoxia stimulates VEGF secretion. Forced overexpression of VEGF in CSCs caused an increase in angiogenesis and cancer formation *in vivo* [80], which further confirmed that CSCs could be a source of enhanced VEGF to stimulate angiogenesis. Folkins *et al.* reported similar data, which found that the CSC high fraction secreted SDF-1, and both VEGF and SDF-1 were crucial for the stimulation of the CSC high fraction on angiogenesis [81]. They found that by inhibiting the VEGF receptor VEGFR2 or the SDF-1 receptor CXCR4, angiogenesis by CM from CSCs was sufficiently blocked. The SDF-1 receptor is highly produced in glioma CSCs, where it activates VEGF secretion *via* the (PI3K)/AKT pathway upon binding to SDF-1 [82]. This shows that SDF-1 has two roles in CSC driven angiogenesis: i) to activated endothelial cells and ii) to gather more VEGF by stimulation of its expression levels in CSCs. MSCs, which activate angiogenesis in prostate cancer [83], interestingly suppress angiogenesis in glioma and its subsequent growth *in vivo* [84]. The expression of pro-angiogenic factors like beg, PDGF-BB and IGF-1 were decreased, suggesting that MSCs suppressed the secretion of these growth factors by glioma cells. However, a study that used MSC-like cells of the glioma showed that MSCs can stimulate angiogenesis [85]. Therefore, the source of the isolated MSCs is a crucial determinant of the imapct of MSCs in glioma [86].

Kidney Cancer

CSCs isolated from renal cancer were associated with pro-angiogenic activities [87]. These CD105 expressing CSCs were able to stimulate angiogenesis by producing exosome-sized micro-vesicles [88], and the CD105-positive micro-vesicles contained RNAs encoding VEGF, an angiogenic factor. The CSC-derived micro-vesicles caused an invasion of vascular endothelial cells, thus protecting them from death and cancer cell adhesion. The lung endothelial cells treatment with the positive micro-vesicles stimulates their expression of the VEGF receptor. Moreover, there is subsequent evidence that the CSC secreted micro- vesicles activate metastasis of kidney cancer cells in the lung.

Colon Cancer

Colon CD133+ CSCs support survival of tumor *via* paracrine mechanisms. The resistance of CD133- non- CSC colon cancer cells to 5-fluorouracil and oxaliplatin decreased when cytokine IL-4 blocks IL-4 antibody [89]. The source of IL-4 traced back to CD133+ CSCs though in colorectal cancer, Th2 lymphocytes, which are a major producer of the inflammatory cytokine have a significant large numbers [90]. CSC derived IL-4 therefore acted as paracrine and autocrine survival factor in colon tumors, where blockage of IL-4 caused suppression of the expression of anti-apoptotic proteins, which suggests that IL-4 may preserve colon cancer cells from cytotoxic agents by stopping apoptosis. Because IL-4 has been shown a protective effect against apoptosis to breast, bladder, prostate and thyroid cancer cells [91, 92], it suggests that IL-4 is an important factor for cancer treatment resistance. Moreover, colon CSCs contributed to the resistance of colon cancer treatment was shown by Emmink *et al.* who compared the secretome of colon CSCs with that of differentiated colon cancer cells [93]. The study demonstrated that CSCs secreted significantly higher levels of ALDH1A1 and BLMH, two enzymes that are known to detoxify chemotherapeutics.

Ovarian Cancer

Cultured ovarian CSCs have been shown to secrete a chemokine CCL5 [94]. This chemokine has been found to be involved in metastasis of breast cancer. CCL5 secretion can be activated by growing breast cancer cells in culture, with mesenchymal stem cells [95]. CCL5 was shown to increase the potential of ovarian CSCs metastasis in an autocrine way but had negligible impact on non-CSC ovarian tumor cells. Nonetheless, invasion of neighboring non-CSC tumor cells was facilitated by secretion of MMP-9 by CSCs, which is caused by the CCL5 feedback loop.

Breast Cancer

Paracrine influences of stem cells in breast cancer have been performed with MSCs, which promote tumor progression through communication with breast cancer cells by way of various soluble factors [96]. Liu *et al.* showed that IL-6 activates MSCs cause the expansion of the CSC pool by a feedback loop where the MSCs secrete chemokine CXCL7, which enhances IL-6 production by breast cancer cells [97]. Adipose derived stem cells were also shown to expand the breast CSC pool by excreting PDGF-D, subsequently inducing epithelial to mesenchymal transition of breast cancer cells and consequent generation of other stem-like cancer cells [98]. There is significant proof that breast CSCs may provide bioactive factors. Transcriptome analyses of CD44$^+$/CD24$^-$ breast CSCs and tumor cells showed an extremely activated TGFβ pathway and plasminogen activator inhibitor-1, TGFβ target gene, which were found to be increased in CSCs. Additionally, PAI-1, an adverse prognostic marker in breast cancer [99], is a excretory protein that activates cell migration and angiogenesis [100, 101]. This protein, produced by MSCs, is able to promote cancer cell migration [102], so it is safe to assume that CSC produced PAI-1 could influence cell movement as well.

Other Cancers

Tumors of the skin (*i.e.* melanoma) have been shown to display a connection between CSCs and endothelial cells. CSCs in skin papillomas secrete significant levels of VEGF and subsequently trigger angiogenesis by promoting VEGFR2 expressing endothelial cells as well as maintain their stem-like nature in an autocrine way *via* the VEGF co-receptor neuropilin 1 [103]. Blocking either receptor in the CSCs or endothelial cells reduce CSC population, which shows that subcutaneous CSCs are proliferated in perivascular areas. These areas are maintained by the VEGF that come directly from the stem cells. CD133 positive melanoma stem cells have been discovered to secrete pro-angiogenic factors like VEGF, and in pancreatic cancer, CD133 positive cancer stem cells produce VEGF-C [104, 105].

CONCLUSION

Although many studies have been focused on the paracrine activity of mesenchymal stem cells in cancer and tissue repair, little is known about the paracrine effects of other adult stem cells and CSCs. Previously, it has been suggested that cancer stem cells and tissue-specific stem cells just function as suppliers of differentiated cells either to compensate the loss of dead cells or to expand the tumor, respectively. Recently, it has been shown that these stem cells regulates both maintenance of tissue and cancer amelioration. For instance, damage-associated molecular pattern (DAMP) following acute kidney failure obviously showed the effect of stem cell-derived factors on tissue recovery [106]. It has been shown that adult stem cells play crucial roles in management of wound repair by producing several factors that provoke angiogenesis and prohibit apoptosis. Both adult stem cells and cancer stems cells secret pro-angiogenic factors, which have particular importance in tissue regeneration and tumor progression, respectively. In addition, stem cells stimulate angiogenesis, which is crucial for cell survival by supplying oxygen and nutrients. Interestingly, endothelial cells are often found close to stem cells. For instance, the hematopoietic stem cells, which reside next to endothelial cells in the bone marrow, are dependent on endothelial cells for their growth [24]. Another example is the glioma stem cells reside in endothelial niches [107]. This positioning is supposed to be of mutual benefit for both glioma stem cells and epithelial cells [108]. Perivascular niches have also been shown regulatory effect on breast cancer cells dormancy [109] and have been found to preserve cancer stem cell populations in skin malignancy [103]. Moreover, endothelial cells probably have a crucial role in cancer metastasis [110]. This similarity between cancer stem cells and endothelial stem cells may provide new insights in the treatment of CSCs which often resistant to chemotherapy [111]. Anti-angiogenic drugs may hinder CSCs from the endothelial feeding layer and cease their growth and differentiation. Nonetheless, testing anti-VEGF in clinical trials for hindering blood and nutrients supply to the tumor has demonstrated insufficient success due to reasons. One of these reasons is the cancer expansion induced by hypoxia [112]. Further investigations is needed to grasp the function of the communication between CSCs and endothelial cells in the progression of cancer to discover new drugs that impede this cell-cell interaction. Collectively, understanding the paracrine influences of stem cells devote to tissue regeneration and cancer progression may aid in finding novel mechanisms of treatment interventions to expedite tissue repair and to enhance cancer therapy, respectively.

CONFLICT OF INTEREST

The authors confirm that they have no conflict of interest to declare for this publication.

ACKNOWLEDGEMENTS

Declared none.

REFERENCES

[1]   Fuchs E, Chen T. A matter of life and death: self-renewal in stem cells. EMBO Rep 2013; 14(1): 39-48.
[http://dx.doi.org/10.1038/embor.2012.197] [PMID: 23229591]

[2]   Hsu YC, Fuchs E. A family business: stem cell progeny join the niche to regulate homeostasis. Nat Rev Mol Cell Biol 2012; 13(2): 103-14.
[http://dx.doi.org/10.1038/nrm3272] [PMID: 22266760]

[3]   Chun Q, Liang LS. Stem cell research, repairing, and regeneration medicine. Int J Low Extrem Wounds 2012; 11(3): 180-3.
[http://dx.doi.org/10.1177/1534734612457035] [PMID: 22885607]

[4]   Fuchs Y, Brown S, Gorenc T, Rodriguez J, Fuchs E, Steller H. Sept4/ARTS regulates stem cell apoptosis and skin regeneration. Science 2013; 341(6143): 286-9.
[http://dx.doi.org/10.1126/science.1233029] [PMID: 23788729]

[5]   Crisostomo PR, Wang M, Markel TA, *et al.* Stem cell mechanisms and paracrine effects: potential in cardiac surgery. Shock 2007; 28(4): 375-83.
[http://dx.doi.org/10.1097/shk.0b013e318058a817] [PMID: 17577135]

[6]   Friedenstein AJ, Piatetzky-Shapiro II, Petrakova KV. Osteogenesis in transplants of bone marrow cells. J Embryol Exp Morphol 1966; 16(3): 381-90.
[PMID: 5336210]

[7]   Zuk PA, Zhu M, Ashjian P, *et al.* Human adipose tissue is a source of multipotent stem cells. Mol Biol Cell 2002; 13(12): 4279-95.
[http://dx.doi.org/10.1091/mbc.E02-02-0105] [PMID: 12475952]

[8]   Chen J, Li Y, Wang L, *et al.* Therapeutic benefit of intravenous administration of bone marrow stromal cells after cerebral ischemia in rats. Stroke 2001; 32(4): 1005-11.
[http://dx.doi.org/10.1161/01.STR.32.4.1005] [PMID: 11283404]

[9]   Ortiz LA, Gambelli F, McBride C, *et al.* Mesenchymal stem cell engraftment in lung is enhanced in response to bleomycin exposure and ameliorates its fibrotic effects. Proc Natl Acad Sci USA 2003; 100(14): 8407-11.
[http://dx.doi.org/10.1073/pnas.1432929100] [PMID: 12815096]

[10]   Park KS, Jung KH, Kim SH, *et al.* Functional expression of ion channels in mesenchymal stem cells derived from umbilical cord vein. Stem Cells 2007; 25(8): 2044-52.
[http://dx.doi.org/10.1634/stemcells.2006-0735] [PMID: 17525238]

[11]   Rojas M, Xu J, Woods CR, *et al.* Bone marrow-derived mesenchymal stem cells in repair of the injured lung. Am J Respir Cell Mol Biol 2005; 33(2): 145-52.
[http://dx.doi.org/10.1165/rcmb.2004-0330OC] [PMID: 15891110]

[12]   Bluguermann C, Wu L, Petrigliano F, McAllister D, Miriuka S, Evseenko DA. Novel aspects of

parenchymal-mesenchymal interactions: from cell types to molecules and beyond. Cell Biochem Funct 2013; 31(4): 271-80.
[http://dx.doi.org/10.1002/cbf.2950] [PMID: 23315627]

[13]   Baraniak PR, McDevitt TC. Stem cell paracrine actions and tissue regeneration. Regen Med 2010; 5(1): 121-43.
[http://dx.doi.org/10.2217/rme.09.74] [PMID: 20017699]

[14]   Kassis I, Vaknin-Dembinsky A, Karussis D. Bone marrow mesenchymal stem cells: agents of immunomodulation and neuroprotection. Curr Stem Cell Res Ther 2011; 6(1): 63-8.
[http://dx.doi.org/10.2174/157488811794480762] [PMID: 20955154]

[15]   Dittmer J, Rody A. Cancer stem cells in breast cancer. Histol Histopathol 2013; 28(7): 827-38.
[PMID: 23468411]

[16]   Dvorak HF. Tumors: wounds that do not heal. Similarities between tumor stroma generation and wound healing. N Engl J Med 1986; 315(26): 1650-9.
[http://dx.doi.org/10.1056/NEJM198612253152606] [PMID: 3537791]

[17]   Kidd S, Spaeth E, Klopp A, Andreeff M, Hall B, Marini FC. The (in) auspicious role of mesenchymal stromal cells in cancer: be it friend or foe. Cytotherapy 2008; 10(7): 657-67.
[http://dx.doi.org/10.1080/14653240802486517] [PMID: 18985472]

[18]   Gupta PB, Fillmore CM, Jiang G, *et al.* Stochastic state transitions give rise to phenotypic equilibrium in populations of cancer cells. Cell 2011; 146(4): 633-44.
[http://dx.doi.org/10.1016/j.cell.2011.07.026] [PMID: 21854987]

[19]   Ren G, Chen X, Dong F, *et al.* Concise review: mesenchymal stem cells and translational medicine: emerging issues. Stem Cells Transl Med 2012; 1(1): 51-8.
[http://dx.doi.org/10.5966/sctm.2011-0019] [PMID: 23197640]

[20]   Cuiffo BG, Karnoub AE. Mesenchymal stem cells in tumor development: emerging roles and concepts. Cell Adhes Migr 2012; 6(3): 220-30.
[http://dx.doi.org/10.4161/cam.20875] [PMID: 22863739]

[21]   Tetta C, Consiglio AL, Bruno S, *et al.* The role of microvesicles derived from mesenchymal stem cells in tissue regeneration; a dream for tendon repair? Muscles Ligaments Tendons J 2012; 2(3): 212-21.
[PMID: 23738299]

[22]   Zimmerlin L, Park TS, Zambidis ET, Donnenberg VS, Donnenberg AD. Mesenchymal stem cell secretome and regenerative therapy after cancer. Biochimie 2013; 95(12): 2235-45.
[http://dx.doi.org/10.1016/j.biochi.2013.05.010] [PMID: 23747841]

[23]   Paul G, Anisimov SV. The secretome of mesenchymal stem cells: potential implications for neuroregeneration. Biochimie 2013; 95(12): 2246-56.
[http://dx.doi.org/10.1016/j.biochi.2013.07.013] [PMID: 23871834]

[24]   Frenette PS, Pinho S, Lucas D, Scheiermann C. Mesenchymal stem cell: keystone of the hematopoietic stem cell niche and a stepping-stone for regenerative medicine. Annu Rev Immunol 2013; 31: 285-316.
[http://dx.doi.org/10.1146/annurev-immunol-032712-095919] [PMID: 23298209]

[25]   Parekkadan B, van Poll D, Suganuma K, *et al.* Mesenchymal stem cell-derived molecules reverse

fulminant hepatic failure. PLoS One 2007; 2(9): e941.
[http://dx.doi.org/10.1371/journal.pone.0000941] [PMID: 17895982]

[26]   Agrawal GK, Jwa NS, Lebrun MH, Job D, Rakwal R. Plant secretome: unlocking secrets of the secreted proteins. Proteomics 2010; 10(4): 799-827.
[http://dx.doi.org/10.1002/pmic.200900514] [PMID: 19953550]

[27]   Camussi G, Deregibus MC, Bruno S, Cantaluppi V, Biancone L. Exosomes/microvesicles as a mechanism of cell-to-cell communication. Kidney Int 2010; 78(9): 838-48.
[http://dx.doi.org/10.1038/ki.2010.278] [PMID: 20703216]

[28]   Muralidharan-Chari V, Clancy JW, Sedgwick A, D ?(tm)Souza-Schorey C. Microvesicles: mediators of extracellular communication during cancer progression. J Cell Sci 2010; 123(Pt 10): 1603-11.
[http://dx.doi.org/10.1242/jcs.064386] [PMID: 20445011]

[29]   Meckes DG Jr, Raab-Traub N. Microvesicles and viral infection. J Virol 2011; 85(24): 12844-54.
[http://dx.doi.org/10.1128/JVI.05853-11] [PMID: 21976651]

[30]   Collino F, Deregibus MC, Bruno S, *et al.* Microvesicles derived from adult human bone marrow and tissue specific mesenchymal stem cells shuttle selected pattern of miRNAs. PLoS One 2010; 5(7): e11803.
[http://dx.doi.org/10.1371/journal.pone.0011803] [PMID: 20668554]

[31]   Yuan A, Farber EL, Rapoport AL, *et al.* Transfer of microRNAs by embryonic stem cell microvesicles. PLoS One 2009; 4(3): e4722.
[http://dx.doi.org/10.1371/journal.pone.0004722] [PMID: 19266099]

[32]   Valadi H, Ekstrom K, Bossios A, Sjostrand M, Lee JJ, Lotvall JO. Exosome-mediated transfer of mRNAs and microRNAs is a novel mechanism of genetic exchange between cells. Nat Cell Biol 2007; 9(6): 654-9.
[http://dx.doi.org/10.1038/ncb1596] [PMID: 17486113]

[33]   Gatti S, Bruno S, Deregibus MC, *et al.* Microvesicles derived from human adult mesenchymal stem cells protect against ischaemia-reperfusion-induced acute and chronic kidney injury. Nephrol Dial Transplant 2011; 26(5): 1474-83.
[http://dx.doi.org/10.1093/ndt/gfr015] [PMID: 21324974]

[34]   Ratajczak J, Miekus K, Kucia M, *et al.* Embryonic stem cell-derived microvesicles reprogram hematopoietic progenitors: evidence for horizontal transfer of mRNA and protein delivery. Leukemia 2006; 20(5): 847-56.
[http://dx.doi.org/10.1038/sj.leu.2404132] [PMID: 16453000]

[35]   Ardoin SP, Shanahan JC, Pisetsky DS. The role of microparticles in inflammation and thrombosis. Scand J Immunol 2007; 66(2-3): 159-65.
[http://dx.doi.org/10.1111/j.1365-3083.2007.01984.x] [PMID: 17635793]

[36]   Huang C, Gu H, Yu Q, Manukyan MC, Poynter JA, Wang M. Sca-1⁺ cardiac stem cells mediate acute cardioprotection *via* paracrine factor SDF-1 following myocardial ischemia/reperfusion. PLoS One 2011; 6(12): e29246.
[http://dx.doi.org/10.1371/journal.pone.0029246] [PMID: 22195033]

[37]   Askari AT, Unzek S, Popovic ZB, *et al.* Effect of stromal-cell-derived factor 1 on stem-cell homing

and tissue regeneration in ischaemic cardiomyopathy. Lancet 2003; 362(9385): 697-703.
[http://dx.doi.org/10.1016/S0140-6736(03)14232-8] [PMID: 12957092]

[38]   Hu X, Dai S, Wu WJ, *et al.* Stromal cell derived factor-1 alpha confers protection against myocardial
        ischemia/reperfusion injury: role of the cardiac stromal cell derived factor-1 alpha CXCR4 axis.
        Circulation 2007; 116(6): 654-63.
        [http://dx.doi.org/10.1161/CIRCULATIONAHA.106.672451] [PMID: 17646584]

[39]   Linke A, MA1/4ller P, Nurzynska D, *et al.* Stem cells in the dog heart are self-renewing, clonogenic,
        and multipotent and regenerate infarcted myocardium, improving cardiac function. Proc Natl Acad Sci
        USA 2005; 102(25): 8966-71.
        [http://dx.doi.org/10.1073/pnas.0502678102] [PMID: 15951423]

[40]   Nesselmann C, Ma N, Bieback K, *et al.* Mesenchymal stem cells and cardiac repair. J Cell Mol Med
        2008; 12(5B): 1795-810.
        [http://dx.doi.org/10.1111/j.1582-4934.2008.00457.x] [PMID: 18684237]

[41]   Noiseux N, Gnecchi M, Lopez-Ilasaca M, *et al.* Mesenchymal stem cells overexpressing Akt
        dramatically repair infarcted myocardium and improve cardiac function despite infrequent cellular
        fusion or differentiation. Mol Ther 2006; 14(6): 840-50.
        [http://dx.doi.org/10.1016/j.ymthe.2006.05.016] [PMID: 16965940]

[42]   Uemura R, Xu M, Ahmad N, Ashraf M. Bone marrow stem cells prevent left ventricular remodeling of
        ischemic heart through paracrine signaling. Circ Res 2006; 98(11): 1414-21.
        [http://dx.doi.org/10.1161/01.RES.0000225952.61196.39] [PMID: 16690882]

[43]   Wang Y, Abarbanell AM, Herrmann JL, *et al.* TLR4 inhibits mesenchymal stem cell (MSC) STAT3
        activation and thereby exerts deleterious effects on MSC-mediated cardioprotection. PLoS One 2010;
        5(12): e14206.
        [http://dx.doi.org/10.1371/journal.pone.0014206] [PMID: 21151968]

[44]   Tang JM, Wang JN, Zhang L, *et al.* VEGF/SDF-1 promotes cardiac stem cell mobilization and
        myocardial repair in the infarcted heart. Cardiovasc Res 2011; 91(3): 402-11.
        [http://dx.doi.org/10.1093/cvr/cvr053] [PMID: 21345805]

[45]   Timmers L, Lim SK, Arslan F, *et al.* Reduction of myocardial infarct size by human mesenchymal
        stem cell conditioned medium. Stem Cell Res (Amst) 2007; 1(2): 129-37.
        [http://dx.doi.org/10.1016/j.scr.2008.02.002] [PMID: 19383393]

[46]   Lai RC, Tan SS, Teh BJ, *et al.* 2012.

[47]   Xu M, Uemura R, Dai Y, Wang Y, Pasha Z, Ashraf M. *In vitro* and *in vivo* effects of bone marrow
        stem cells on cardiac structure and function. J Mol Cell Cardiol 2007; 42(2): 441-8.
        [http://dx.doi.org/10.1016/j.yjmcc.2006.10.009] [PMID: 17187821]

[48]   Timmers L, Lim SK, Hoefer IE, *et al.* Human mesenchymal stem cell-conditioned medium improves
        cardiac function following myocardial infarction. Stem Cell Res (Amst) 2011; 6(3): 206-14.
        [http://dx.doi.org/10.1016/j.scr.2011.01.001] [PMID: 21419744]

[49]   Gnecchi M, He H, Noiseux N, *et al.* Evidence supporting paracrine hypothesis for Akt-modified
        mesenchymal stem cell-mediated cardiac protection and functional improvement. FASEB J 2006;
        20(6): 661-9.

[http://dx.doi.org/10.1096/fj.05-5211com] [PMID: 16581974]

[50]   Kinnaird T, Stabile E, Burnett MS, *et al.* Local delivery of marrow-derived stromal cells augments collateral perfusion through paracrine mechanisms. Circulation 2004; 109(12): 1543-9.
[http://dx.doi.org/10.1161/01.CIR.0000124062.31102.57] [PMID: 15023891]

[51]   Estrada R, Li N, Sarojini H, An J, Lee MJ, Wang E. Secretome from mesenchymal stem cells induces angiogenesis *via* Cyr61. J Cell Physiol 2009; 219(3): 563-71.
[http://dx.doi.org/10.1002/jcp.21701] [PMID: 19170074]

[52]   Ohnishi S, Sumiyoshi H, Kitamura S, Nagaya N. Mesenchymal stem cells attenuate cardiac fibroblast proliferation and collagen synthesis through paracrine actions. FEBS Lett 2007; 581(21): 3961-6.
[http://dx.doi.org/10.1016/j.febslet.2007.07.028] [PMID: 17662720]

[53]   Xu X, Xu Z, Xu Y, Cui G. Effects of mesenchymal stem cell transplantation on extracellular matrix after myocardial infarction in rats. Coron Artery Dis 2005; 16(4): 245-55.
[http://dx.doi.org/10.1097/00019501-200506000-00006] [PMID: 15915077]

[54]   Cho HJ, Lee N, Lee JY, *et al.* Role of host tissues for sustained humoral effects after endothelial progenitor cell transplantation into the ischemic heart. J Exp Med 2007; 204(13): 3257-69.
[http://dx.doi.org/10.1084/jem.20070166] [PMID: 18070934]

[55]   Oshima H, Payne TR, Urish KL, *et al.* Differential myocardial infarct repair with muscle stem cells compared to myoblasts. Mol Ther 2005; 12(6): 1130-41.
[http://dx.doi.org/10.1016/j.ymthe.2005.07.686] [PMID: 16125468]

[56]   Kupatt C, Bock-Marquette I, Boekstegers P. Embryonic endothelial progenitor cell-mediated cardioprotection requires Thymosin beta4. Trends Cardiovasc Med 2008; 18(6): 205-10.
[http://dx.doi.org/10.1016/j.tcm.2008.10.002] [PMID: 19185810]

[57]   Payne TR, Oshima H, Okada M, *et al.* A relationship between vascular endothelial growth factor, angiogenesis, and cardiac repair after muscle stem cell transplantation into ischemic hearts. J Am Coll Cardiol 2007; 50(17): 1677-84.
[http://dx.doi.org/10.1016/j.jacc.2007.04.100] [PMID: 17950150]

[58]   Wu G, Rana JS, Wykrzykowska J, *et al.* Exercise-induced expression of VEGF and salvation of myocardium in the early stage of myocardial infarction. Am J Physiol Heart Circ Physiol 2009; 296(2): H389-95.
[http://dx.doi.org/10.1152/ajpheart.01393.2007] [PMID: 19060119]

[59]   Ambrosio F, Wolf SL, Delitto A, *et al.* The emerging relationship between regenerative medicine and physical therapeutics. Phys Ther 2010; 90(12): 1807-14.
[http://dx.doi.org/10.2522/ptj.20100030] [PMID: 21030663]

[60]   Li TS, Cheng K, Malliaras K, *et al.* Direct comparison of different stem cell types and subpopulations reveals superior paracrine potency and myocardial repair efficacy with cardiosphere-derived cells. J Am Coll Cardiol 2012; 59(10): 942-53.
[http://dx.doi.org/10.1016/j.jacc.2011.11.029] [PMID: 22381431]

[61]   Duran JM, Makarewich CA, Sharp TE, *et al.* Bone-derived stem cells repair the heart after myocardial infarction through transdifferentiation and paracrine signaling mechanisms. Circ Res 2013; 113(5): 539-52.

[http://dx.doi.org/10.1161/CIRCRESAHA.113.301202] [PMID: 23801066]

[62]   Horie N, Pereira MP, Niizuma K, *et al.* Transplanted stem cell-secreted vascular endothelial growth factor effects poststroke recovery, inflammation, and vascular repair. Stem Cells 2011; 29(2): 274-85.
[http://dx.doi.org/10.1002/stem.584] [PMID: 21732485]

[63]   Andres RH, Horie N, Slikker W, *et al.* Human neural stem cells enhance structural plasticity and axonal transport in the ischaemic brain. Brain 2011; 134(Pt 6): 1777-89.
[http://dx.doi.org/10.1093/brain/awr094] [PMID: 21616972]

[64]   Liauw J, Hoang S, Choi M, *et al.* Thrombospondins 1 and 2 are necessary for synaptic plasticity and functional recovery after stroke. J Cereb Blood Flow Metab 2008; 28(10): 1722-32.
[http://dx.doi.org/10.1038/jcbfm.2008.65] [PMID: 18594557]

[65]   Lu P, Jones LL, Snyder EY, Tuszynski MH. Neural stem cells constitutively secrete neurotrophic factors and promote extensive host axonal growth after spinal cord injury. Exp Neurol 2003; 181(2): 115-29.
[http://dx.doi.org/10.1016/S0014-4886(03)00037-2] [PMID: 12781986]

[66]   Cantinieaux D, Quertainmont R, Blacher S, *et al.* Conditioned medium from bone marrow-derived mesenchymal stem cells improves recovery after spinal cord injury in rats: an original strategy to avoid cell transplantation. PLoS One 2013; 8(8): e69515.
[http://dx.doi.org/10.1371/journal.pone.0069515] [PMID: 24013448]

[67]   Crigler L, Robey RC, Asawachaicharn A, Gaupp D, Phinney DG. Human mesenchymal stem cell subpopulations express a variety of neuro-regulatory molecules and promote neuronal cell survival and neuritogenesis. Exp Neurol 2006; 198(1): 54-64.
[http://dx.doi.org/10.1016/j.expneurol.2005.10.029] [PMID: 16336965]

[68]   Sallustio F, Costantino V, Cox SN, *et al.* Human renal stem/progenitor cells repair tubular epithelial cell injury through TLR2-driven inhibin-A and microvesicle-shuttled decorin. Kidney Int 2013; 83(3): 392-403.
[http://dx.doi.org/10.1038/ki.2012.413] [PMID: 23325086]

[69]   Maeshima A, Zhang YQ, Furukawa M, Naruse T, Kojima I. Hepatocyte growth factor induces branching tubulogenesis in MDCK cells by modulating the activin-follistatin system. Kidney Int 2000; 58(4): 1511-22.
[http://dx.doi.org/10.1046/j.1523-1755.2000.00313.x] [PMID: 11012886]

[70]   Togel F, Weiss K, Yang Y, Hu Z, Zhang P, Westenfelder C. Vasculotropic, paracrine actions of infused mesenchymal stem cells are important to the recovery from acute kidney injury. Am J Physiol Renal Physiol 2007; 292(5): F1626-35.
[http://dx.doi.org/10.1152/ajprenal.00339.2006] [PMID: 17213465]

[71]   Imberti B, Morigi M, Tomasoni S, *et al.* Insulin-like growth factor-1 sustains stem cell mediated renal repair. J Am Soc Nephrol 2007; 18(11): 2921-8.
[http://dx.doi.org/10.1681/ASN.2006121318] [PMID: 17942965]

[72]   van Koppen A, Joles JA, van Balkom BW, *et al.* Human embryonic mesenchymal stem cell-derived conditioned medium rescues kidney function in rats with established chronic kidney disease. PLoS One 2012; 7(6): e38746.
[http://dx.doi.org/10.1371/journal.pone.0038746] [PMID: 22723882]

[73]   Mintz PJ, Huang KW, Reebye V, *et al.* Exploiting human CD34+ stem cell-conditioned medium for tissue repair. Mol Ther 2014; 22(1): 149-59.
[PMID: 23985698]

[74]   Hogaboam CM, Bone-Larson CL, Steinhauser ML, *et al.* Novel CXCR2-dependent liver regenerative qualities of ELR-containing CXC chemokines. FASEB J 1999; 13(12): 1565-74.
[PMID: 10463948]

[75]   Wei J, Barr J, Kong LY, *et al.* Glioblastoma cancer-initiating cells inhibit T-cell proliferation and effector responses by the signal transducers and activators of transcription 3 pathway. Mol Cancer Ther 2010; 9(1): 67-78.
[http://dx.doi.org/10.1158/1535-7163.MCT-09-0734] [PMID: 20053772]

[76]   Wei J, Barr J, Kong LY, *et al.* Glioma-associated cancer-initiating cells induce immunosuppression. Clin Cancer Res 2010; 16(2): 461-73.
[http://dx.doi.org/10.1158/1078-0432.CCR-09-1983] [PMID: 20068105]

[77]   Peng W, Wang HY, Miyahara Y, Peng G, Wang RF. Tumor-associated galectin-3 modulates the function of tumor-reactive T cells. Cancer Res 2008; 68(17): 7228-36.
[http://dx.doi.org/10.1158/0008-5472.CAN-08-1245] [PMID: 18757439]

[78]   Kuklinski S, Pesheva P, Heimann C, *et al.* Expression pattern of galectin-3 in neural tumor cell lines. J Neurosci Res 2000; 60(1): 45-57.
[http://dx.doi.org/10.1002/(SICI)1097-4547(20000401)60:1<45::AID-JNR5>3.0.CO;2-Y]   [PMID: 10723067]

[79]   Bao S, Wu Q, Sathornsumetee S, *et al.* Stem cell-like glioma cells promote tumor angiogenesis through vascular endothelial growth factor. Cancer Res 2006; 66(16): 7843-8.
[http://dx.doi.org/10.1158/0008-5472.CAN-06-1010] [PMID: 16912155]

[80]   Oka N, Soeda A, Inagaki A, *et al.* VEGF promotes tumorigenesis and angiogenesis of human glioblastoma stem cells. Biochem Biophys Res Commun 2007; 360(3): 553-9.
[http://dx.doi.org/10.1016/j.bbrc.2007.06.094] [PMID: 17618600]

[81]   Folkins C, Shaked Y, Man S, *et al.* Glioma tumor stem-like cells promote tumor angiogenesis and vasculogenesis *via* vascular endothelial growth factor and stromal-derived factor 1. Cancer Res 2009; 69(18): 7243-51.
[http://dx.doi.org/10.1158/0008-5472.CAN-09-0167] [PMID: 19738068]

[82]   Ping YF, Yao XH, Jiang JY, *et al.* The chemokine CXCL12 and its receptor CXCR4 promote glioma stem cell-mediated VEGF production and tumour angiogenesis *via* PI3K/AKT signalling. J Pathol 2011; 224(3): 344-54.
[http://dx.doi.org/10.1002/path.2908] [PMID: 21618540]

[83]   Lin G, Yang R, Banie L, *et al.* Effects of transplantation of adipose tissue-derived stem cells on prostate tumor. Prostate 2010; 70(10): 1066-73.
[http://dx.doi.org/10.1002/pros.21140] [PMID: 20232361]

[84]   Ho IA, Toh HC, Ng WH, *et al.* Human bone marrow-derived mesenchymal stem cells suppress human glioma growth through inhibition of angiogenesis. Stem Cells 2013; 31(1): 146-55.
[http://dx.doi.org/10.1002/stem.1247] [PMID: 23034897]

[85]   Kong BH, Shin HD, Kim SH, *et al.* Increased *in vivo* angiogenic effect of glioma stromal mesenchymal stem-like cells on glioma cancer stem cells from patients with glioblastoma. Int J Oncol 2013; 42(5): 1754-62.
[PMID: 23483121]

[86]   Akimoto K, Kimura K, Nagano M, *et al.* Umbilical cord blood-derived mesenchymal stem cells inhibit, but adipose tissue-derived mesenchymal stem cells promote, glioblastoma multiforme proliferation. Stem Cells Dev 2013; 22(9): 1370-86.
[http://dx.doi.org/10.1089/scd.2012.0486] [PMID: 23231075]

[87]   Bussolati B, Dekel B, Azzarone B, Camussi G. Human renal cancer stem cells. Cancer Lett 2013; 338(1): 141-6.
[http://dx.doi.org/10.1016/j.canlet.2012.05.007] [PMID: 22587951]

[88]   Grange C, Tapparo M, Collino F, *et al.* Microvesicles released from human renal cancer stem cells stimulate angiogenesis and formation of lung premetastatic niche. Cancer Res 2011; 71(15): 5346-56.
[http://dx.doi.org/10.1158/0008-5472.CAN-11-0241] [PMID: 21670082]

[89]   Todaro M, Alea MP, Di Stefano AB, *et al.* Colon cancer stem cells dictate tumor growth and resist cell death by production of interleukin-4. Cell Stem Cell 2007; 1(4): 389-402.
[http://dx.doi.org/10.1016/j.stem.2007.08.001] [PMID: 18371377]

[90]   Francipane MG, Alea MP, Lombardo Y, Todaro M, Medema JP, Stassi G. Crucial role of interleukin-4 in the survival of colon cancer stem cells. Cancer Res 2008; 68(11): 4022-5.
[http://dx.doi.org/10.1158/0008-5472.CAN-07-6874] [PMID: 18519657]

[91]   Conticello C, Pedini F, Zeuner A, *et al.* IL-4 protects tumor cells from anti-CD95 and chemotherapeutic agents *via* up-regulation of antiapoptotic proteins. J Immunol 2004; 172(9): 5467-77.
[http://dx.doi.org/10.4049/jimmunol.172.9.5467] [PMID: 15100288]

[92]   Todaro M, Zerilli M, Ricci-Vitiani L, *et al.* Autocrine production of interleukin-4 and interleukin-10 is required for survival and growth of thyroid cancer cells. Cancer Res 2006; 66(3): 1491-9.
[http://dx.doi.org/10.1158/0008-5472.CAN-05-2514] [PMID: 16452205]

[93]   Emmink BL, Verheem A, Van Houdt WJ, *et al.* The secretome of colon cancer stem cells contains drug-metabolizing enzymes. J Proteomics 2013; 91: 84-96.
[http://dx.doi.org/10.1016/j.jprot.2013.06.027] [PMID: 23835434]

[94]   Long H, Xie R, Xiang T, *et al.* Autocrine CCL5 signaling promotes invasion and migration of CD133+ ovarian cancer stem-like cells via NF-I B-mediated MMP-9 upregulation. Stem Cells 2012; 30(10): 2309-19.
[http://dx.doi.org/10.1002/stem.1194] [PMID: 22887854]

[95]   Karnoub AE, Dash AB, Vo AP, *et al.* Mesenchymal stem cells within tumour stroma promote breast cancer metastasis. Nature 2007; 449(7162): 557-63.
[http://dx.doi.org/10.1038/nature06188] [PMID: 17914389]

[96]   Dittmer J, Oerlecke I, Leyh B. Involvement of mesenchymal stem cells in breast cancer progression. In: Gunduz M, Gunduz E, Eds. Breast Cancer-Focusing Tumor Microenvironment, Stem Cells and Metastasis. Rijeka: INTECH Open Access Publisher 2011; pp. 247-72.

[http://dx.doi.org/10.5772/21325]

[97] Liu S, Ginestier C, Ou SJ, *et al.* Breast cancer stem cells are regulated by mesenchymal stem cells through cytokine networks. Cancer Res 2011; 71(2): 614-24.
[http://dx.doi.org/10.1158/0008-5472.CAN-10-0538] [PMID: 21224357]

[98] Devarajan E, Song YH, Krishnappa S, Alt E. Epithelial-mesenchymal transition in breast cancer lines is mediated through PDGF-D released by tissue-resident stem cells. Int J Cancer 2012; 131(5): 1023-31.
[http://dx.doi.org/10.1002/ijc.26493] [PMID: 22038895]

[99] Harbeck N, Schmitt M, Meisner C, *et al.* Chemo-N 0 Study Group. Ten-year analysis of the prospective multicentre Chemo-N0 trial validates American Society of Clinical Oncology (ASCO)-recommended biomarkers uPA and PAI-1 for therapy decision making in node-negative breast cancer patients. Eur J Cancer 2013; 49(8): 1825-35.
[http://dx.doi.org/10.1016/j.ejca.2013.01.007] [PMID: 23490655]

[100] Dellas C, Loskutoff DJ. Historical analysis of PAI-1 from its discovery to its potential role in cell motility and disease. Thromb Haemost 2005; 93(4): 631-40.
[PMID: 15841306]

[101] Czekay RP, Wilkins-Port CE, Higgins SP, *et al.* 2011.

[102] Hogan NM, Joyce MR, Murphy JM, *et al.* Impact of mesenchymal stem cell secreted PAI-1 on colon cancer cell migration and proliferation. Biochem Biophys Res Commun 2013; 435(4): 574-9.
[http://dx.doi.org/10.1016/j.bbrc.2013.05.013] [PMID: 23685140]

[103] Beck B, Driessens G, Goossens S, *et al.* A vascular niche and a VEGF-Nrp1 loop regulate the initiation and stemness of skin tumours. Nature 2011; 478(7369): 399-403.
[http://dx.doi.org/10.1038/nature10525] [PMID: 22012397]

[104] Monzani E, Facchetti F, Galmozzi E, *et al.* Melanoma contains CD133 and ABCG2 positive cells with enhanced tumourigenic potential. Eur J Cancer 2007; 43(5): 935-46.
[http://dx.doi.org/10.1016/j.ejca.2007.01.017] [PMID: 17320377]

[105] Maeda S, Shinchi H, Kurahara H, *et al.* CD133 expression is correlated with lymph node metastasis and vascular endothelial growth factor-C expression in pancreatic cancer. Br J Cancer 2008; 98(8): 1389-97.
[http://dx.doi.org/10.1038/sj.bjc.6604307] [PMID: 18349830]

[106] Romagnani P, Anders HJ. What can tubular progenitor cultures teach us about kidney regeneration? Kidney Int 2013; 83(3): 351-3.
[http://dx.doi.org/10.1038/ki.2012.437] [PMID: 23446253]

[107] Calabrese C, Poppleton H, Kocak M, *et al.* A perivascular niche for brain tumor stem cells. Cancer Cell 2007; 11(1): 69-82.
[http://dx.doi.org/10.1016/j.ccr.2006.11.020] [PMID: 17222791]

[108] Eyler CE, Rich JN. Survival of the fittest: cancer stem cells in therapeutic resistance and angiogenesis. J Clin Oncol 2008; 26(17): 2839-45.
[http://dx.doi.org/10.1200/JCO.2007.15.1829] [PMID: 18539962]

[109] Ghajar CM, Peinado H, Mori H, *et al.* The perivascular niche regulates breast tumour dormancy. Nat

Cell Biol 2013; 15(7): 807-17.
[http://dx.doi.org/10.1038/ncb2767] [PMID: 23728425]

[110] Descot A, Oskarsson T. The molecular composition of the metastatic niche. Exp Cell Res 2013; 319(11): 1679-86.
[http://dx.doi.org/10.1016/j.yexcr.2013.04.017] [PMID: 23707205]

[111] Gupta PB, Onder TT, Jiang G, *et al.* Identification of selective inhibitors of cancer stem cells by high-throughput screening. Cell 2009; 138(4): 645-59.
[http://dx.doi.org/10.1016/j.cell.2009.06.034] [PMID: 19682730]

[112] Bottos A, Bardelli A. Oncogenes and angiogenesis: a way to personalize anti-angiogenic therapy? Cell Mol Life Sci 2013; 70(21): 4131-40.
[http://dx.doi.org/10.1007/s00018-013-1331-3] [PMID: 23685900]

CHAPTER 8

Stem Cells in Cancer Development and Therapy

Azza El Amir[*] and Wajeet Nabil[*]

Zoology Department, Faculty of Science, Cairo University, Egypt

Abstract: Tumorigenic cancer stem cells [CSCs] are multipotent cells that found together with their nontumorigenic variants in the same clone. Although sharing the same genetic battery, CSCs and their non-tumorigenic progenies respond differently to micro environmental stresses. Thus, hypoxia, inflammation, low pH, shortage in nutrients and cancer therapies result in broad spectrum of changes in signaling pathways. These facts paved the road to a new era in cancer treatment strategies where inhibiting CSCs specific pathways are combined with traditional therapies of bulk tumor cells. Despite the promise such combination brought to cancer cure improvement, resistance to some to some CSCs-related therapies is noticed. Thus, this review focuses on the mechanisms that may involve in such resistance including drug efflux by ABC transporters, activation of aldehyde dehydrogenase and developmental pathways, enhanced DNAdamage response and autophagy and microenvironmental conditions. This review discusses as well the possible therapeutic strategies for improving cancer treatment.

Keywords: Carcinogenesis, Cancer therapeutic strategies, Cell surface markers, Inflammatory reactions, Nanotechnology, Role of niche, Tumorigenic cancer stem cells [CSCs].

MAY CANCER BE A DISEASE OF STEM CELLS?

Based on the fact of sharing the characteristic of uncontrolled proliferation among almost all cancer types, two therapeutic approaches; differentiation [1] and destructive [2] therapies; for combating cancer have been developed.

[*] **Correspondence authors Azza El Amir and Wajeet Nabil:** Zoology Department, Faculty of Science, Cairo University, Egypt; E-mail: wajeetnabil@yahoo.com; E-mail: azzaelamir@yahoo.com

Ahmed El-Hashash (Ed)

Although these approaches aimed either to induce differentiation or interrupt the malignant proliferation, they cannot completely cure cancer.

Upon clinical perceptions and hereditary investigations of a mixed bag of growth, the six-hereditary transformation theory has raised the mid-1990s. This theory proposed that for the transformation of an ordinary substantial cell into a tumor cell, six transformations are needed [3, 4]. These transformations incorporate (a) independence for development signals, (b) lack of care to antigrowth signals, (c) avoidance of apoptosis, (d) boundless capacity to recreate, (e) supported angiogenesis, and (f) tissue attack and metastasis. The stem cells; especially cancer stem cells; have the most astounding expansion potential and a longer life compass than their progeny, they have great susceptibility for accumulation of genetic mutations [5].

As well the presence of stem cells, even in small numbers, in the adult body highlighted the possible role of stem cells in initiating cancer. The history of such theory comes back to the concept of "the embryonal rest" in the 19th century [6, 7]. Recently, convincing proof for the stem cell concept of cancerogenesis was first demonstrated by study of Dick and coworkers in 1994 showed that only minor population of leukemia cells express specific surface markers and are responsible for tumorigenicity [8, 9]. In 2007, cancer stem cells were proven by Michael Clarke's lab to be responsible for its tumorigenic and metastatic ability despite of their presence in only minor fraction [10, 11]. A fact that may explains the failure of some treatments to completely cure some types of cancer. Such treatments reduce the tumor mass by removing the proliferating cells where cancer stem cells escaped that scenario due to their slow cycle. Thus limited resemblance in the profiles of gene expression between samples of heterogeneous tumor tissue and tumor stem cell population should be expected [12, 13].

Nevertheless, cancer remains a major health problem in many regions of the world causing about 20% of all deaths in developed countries. The first investigations of the intratumor heterogeneity demonstrated that a high degree of genetic instability in melanoma cancer cells corresponds to a higher rate of generation of cell clones resistant to the chemo-therapeutic drugs methotrexate and N-phosphonacetyl-l- aspartames [14]. This may present a principal challenge to the targeted therapy. Progression in cancer treatment and prevention has achieved by the validation of new options for diagnosis and treatment through the multidisciplinary cancer research [15]. Thus understanding the yet unexplored biology and environment of CSCs will lead to a breakthrough in therapeutic innovations.

Thus in conclusion, accumulating evidence coming from a number of whole genome sequencing studies shattered a widely accepted concept that tumor arises from one single clonogenic cell which accumulates multiple mutations in a stepwise manner. In addition, recent study revealed the co-existence of multiple genetically diverse clones within the same tumor [16 - 21]. Different clones may exhibit distinct mechanisms of resistances within the same tumor due to genetic heterogeneity. Also, there are fundamental functional and phenotypic differences between the same clone cells.

DEVELOPMENT OF STEM CELLS

Although normal adult somatic stem cells and cancer stem cells are similar in their ability of self-renewal and slow cycling, their destinations are distinctive. This fact suggests a controlling role of the niche under normal conditions. If such a role is valid for cancer stem cells or not was a matter of question. It is well known that mutations in stem/progenitor cells result in excessively proliferation cancer stem cells while the role of micro environmental growth promoting signals in initiation of cancer stem cells was just suggested. The presence of certain types of tumors that arise as a result of coordination of mutations of two or more different types of cells *e.g.* certain types of neuroblastoma the first evidence for that suggestion [22]. More facts including the *de novo* angiogenesis for rapid tumor growth and molecular machinery used by cancer cells for invasion and metastasis are convincing proof for the potential role of the niche Well understanding of the mechanisms of the suspected potential role of the niche will be promising in better understanding of cancer cell metastasis and will help in developing treatment that will specifically target cancer stem cells with any adverse effects on the normal stem cells self renewal (Fig.**1**).

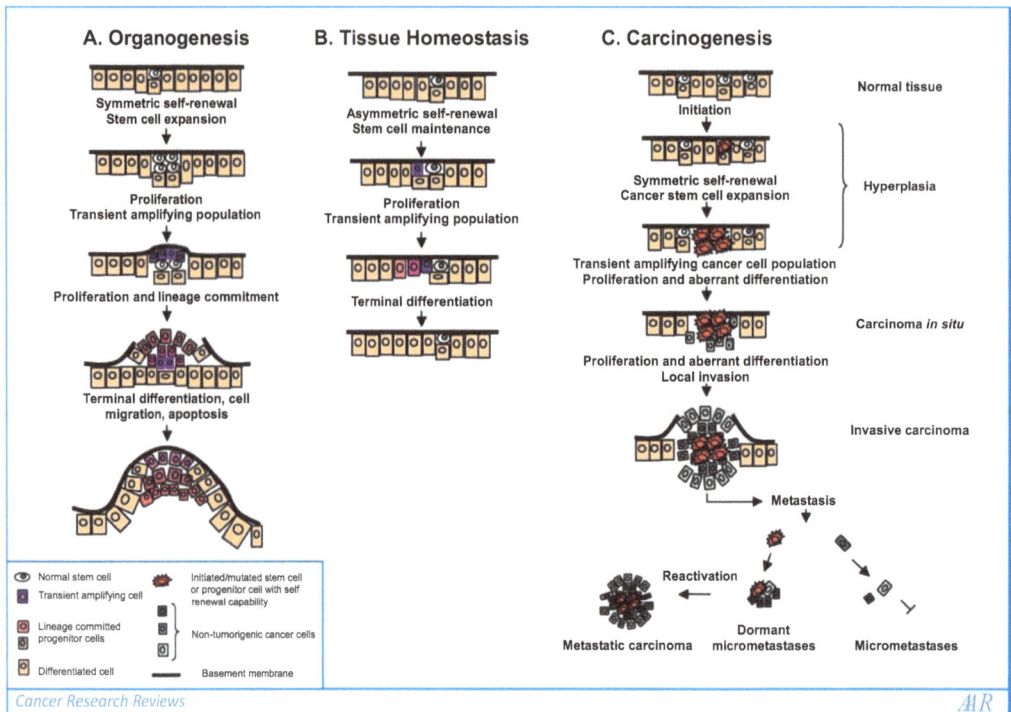

Fig. (1). Stem cells in health and carcinogenesis. **A:** niche controlled stem cell expansion due to normal development and symmetric self-renewal. Stem cells pass through a long way starting by differentiation into a transient amplifying population and undergo lineage establishment passing through cells migration and terminal cell differentiation till ending in apoptosis of differentiated cells **B:** typical tissue homeostasis established by maintenance of stems cell as a result of stem cell asymmetric self-renewal. Dividing and achieving function of transient amplifying progenitor cells compensate normal cell loss. **C:** Carcinogenesis: cancer stem cells expansion results from dysregulated symmetric self-renewal whereas, invasion and metastasis may result from further epigenetic changes or mutations. It was observed that the tumor recurrence is mainly due to reactivation of previously disseminated and dormant cancer stem cells, whereas dissemination of differentiated tumor cell has limited recurrence effect [23].

A great importance of the cancer stem model in cancer risk reduction, early detection, prevention and treatment has evolved (Fig. **2**). Cancer risk may be decreased if the number of the normal stem cells is reduced, while factors secreted by initiated stem cells may be of great benefit for the early diagnosis of cancers. As well, induction of apoptosis of differentiation of initiated stem cells has a great implication on cancer prevention [23]. Normally, the stem cell niche provides only transient proliferating signal necessary for tissue regeneration. In cancer or tumor, cancer stem cells revolt against the control of the niche and undergo uncontrolled self dependent proliferation. Sometimes, the niche has dominant growth promoting signal. Combination of these two pathways is possible (Fig. **3**) [24].

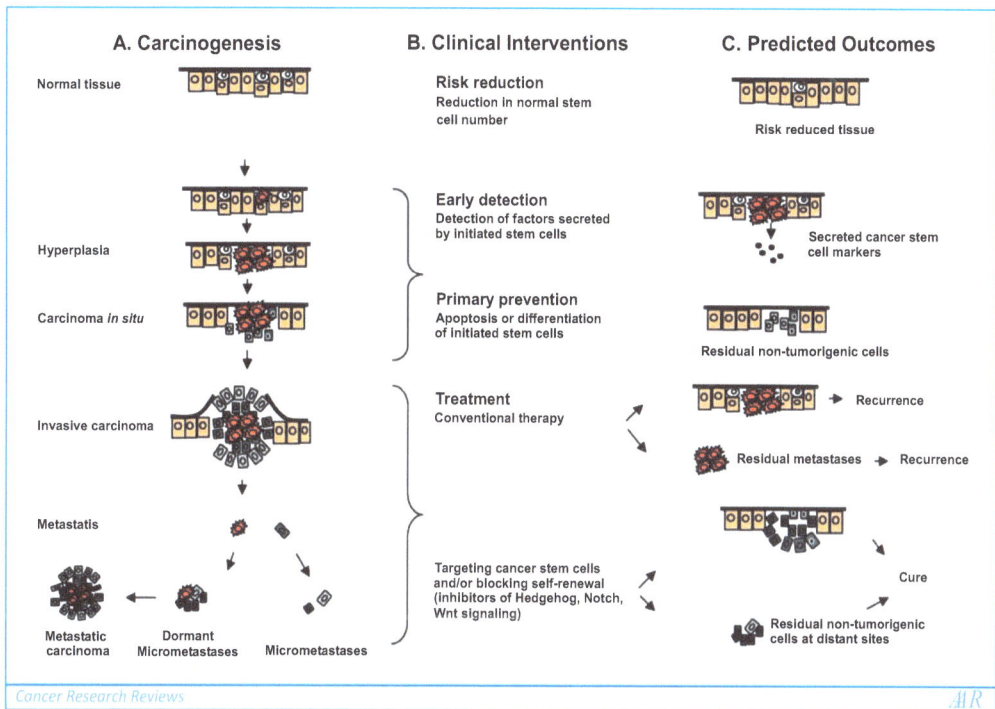

Fig. (2). Importance of cancer stem cell model in carcinogenesis.

EFFECTS OF INFLAMMATORY REACTIONS ON STEM CELLS

Inflammatory events have great importance when exploring cancer biology and stem cell therapy. Inflammation encompasses diverse molecular pathway, the most interesting nowadays is how the mediators of the inflammation interact with both stem cells and cancer cells. Cytokines could be the heroes in the battle against the various types of malignancies. The cell-based therapy in carcinogenesis significantly relies on the IL-2 mediated tumor cell lysis. As well the presence of integrin antagonists is essential because of their ability to regulate differentiation, survival and angiogenesis. Although evidences show the possibility of maximizing the effect of chemotherapy and radiotherapy by using integrin antagonists, integrins of each cancer type should be specifically targeted which is an extraordinarily difficult mission. Caution must be taken or patient will be harmed. Also, the nuclear factor-kappa B [NFκB] is so important as it is central to inflammatory response and is the link between inflammation and stem-cells. Mesenchymal stem cells, with their anti-inflammatory and immunosuppressive effects, are superior in facing the challenges in regenerative medicine. They provide a great potential for therapy when significant inflammatory component is faced, as well in graft-*versus* host disease and tissue regeneration [25].

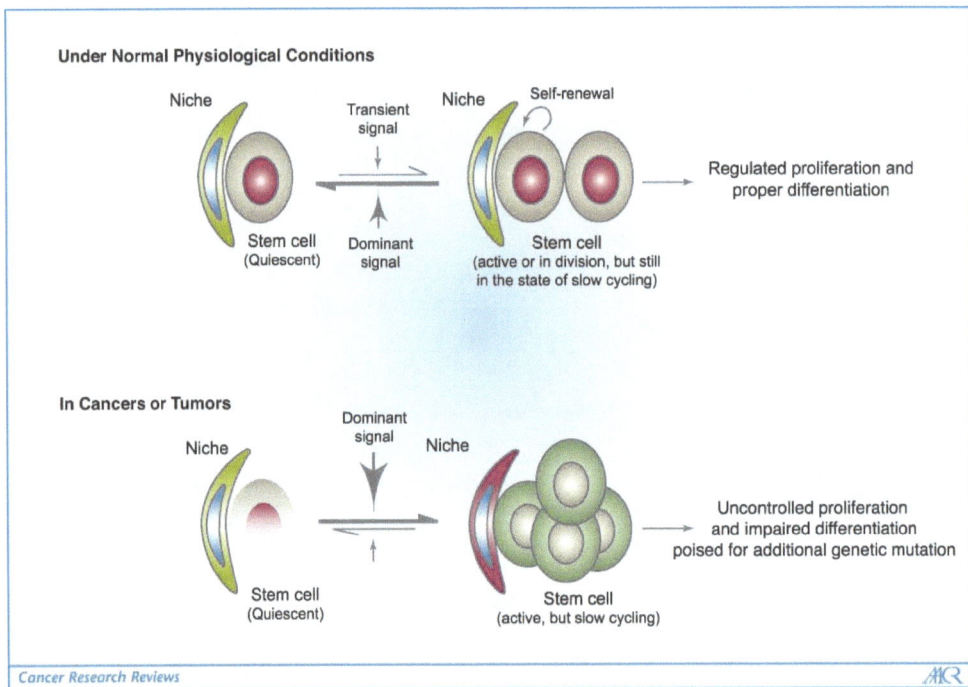

Fig. (3). The suggested role of niche in carcinogenesis.

THE DRIFT OF NORMAL STEM CELLS INTO CANCR STEM CELLS

In the field of cancer with its highest mortality rate worldwide, stem cells gain their importance from the accumulating evidences supporting the "cancer stem model" of carcinogenesis. Cancer stem cells are pointed to as the main cause of malignancy, growth and recurrence. That brings the stem cells in focus in every trial for either prevention or therapy of cancers especially those that are incurable [26]. Likewise, chemotherapeutics are practically ineffective in hepatic carcinoma, with a median response below 15% [27]. Clearly, increased understanding of cancer mechanisms, especially the biology of therapeutic resistance and clinically relevant patient- specific molecular variation in cancers will be crucial for improving outcomes with better therapies. Interpreting the Mystery New research suggests the capability of cancer stem cells in recapitulating the cancer *via* initiation and propagation as only a few number of them [100 cells] can initiate tumors with high efficiency [28, 29]. Cancer stem cells still have a debated origin; they may result from oncogenic mutation of normal stem cells, or from adult neoplastic cells with stem cell properties *via* mutations [30]. This model challenges the assumptions of the stochastic theory and suggests that re- examination of cancer mechanisms and cell-targeted therapies would be informative for future therapies. Analysis of specific CSC-related biomarkers might be important for tumor diagnostics and staging, for patient prognosis, prediction of patients' response to therapies and for treatment selection. CSC- specific therapies targeting CSC surface markers, CSC niche, signaling mechanisms playing a role in the maintenance and survival of CSC and inhibition of CSC metabolic pathways can be combined with conventional radio- and chemotherapies and could be beneficial in improving cancer treatment [31].

CANCER STEM CELL HYPOTHESIS

Recently, the various implications of stem cells in the field of cancer gained great interest due to the new facilities provided to measure self-renewal (Fig. **4**). The long-living nature of stem cells makes them a hotspot of accumulation of carcinogenic mutations. A clear evidence of cancer development due to accumulation of mutations over a long period of time is seen among women that exposed to atomic bomb radiation in Hiroshima. They developed breast cancer 20-30 years after exposure [32]. Further evidences of the "accumulation of mutations" hypothesis, as a cause of cancerogenesis, are revealed. Women in late adolescence have the greatest risk of developing breast cancer upon exposure to radiation as stem cells are at their highest number in the mammary gland during this period [33].

What makes the stem cells a major suspect in the carcinogenesis is the common properties they share with cancer cells. The capability of self-renewal, active telomerase expression, activation of anti- apoptotic pathway, increased activity of membrane transporters and the ability to migrate and metastasize are the most important.

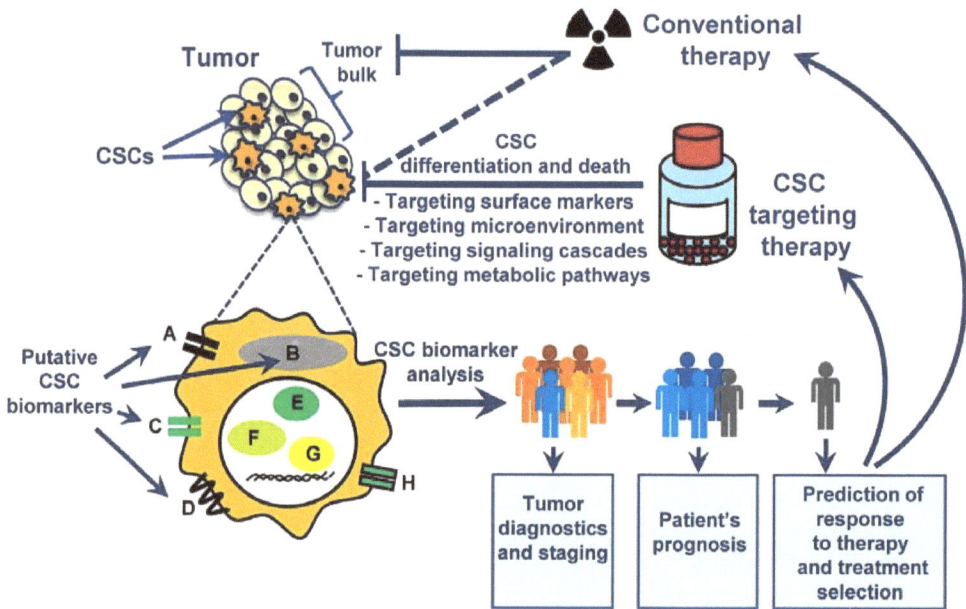

Fig. (4). Future perspectives for development of combination therapies targeting bulk tumor cells and CSC population.

Dysregulation of the self-renewal process may be the early event in transformation due to shuffling from asymmetric into symmetric division during stem cell expansion (Fig. **3**) [34]. A role of the signals produced from the niche of the stem cell in controlling the self-renewal process is proved as only single hematopoietic stem cells is able to reconstitute the whole hematopoietic system by extensive cell expansion. This process stops when the pool is replenished.

The self-renewal of normal stem cells is regulated by specific signaling pathways. The implication of Wnt, Notch, and Hedgehog signaling pathways in the regulation of self- renewal of mammary, hematopoietic and neuronal stem cells is elucidated [34, 35]. Dysregulation of these pathways has proved to have a tumorigenic effect in both rodent models and human. Evidences for association between certain signaling pathway defects and specific types of cancers have shown. Early in colon cancer carcinogenesis, defects in the Wnt signaling pathway are seen. Whereas defects in Hedgehog signaling were first noticed in human basal carcinomas of the skin [36], recent studies show more evidences of the involvement of that pathway in human pancreatic, gastric, prostate and breast carcinoma [37, 38].

In human T-cell acute lymphoblastic leukemia, cervical cancer and breast cancer, alteration in Notch signaling were observed [39 - 43]. Recently, studies suggested that alteration in the signaling pathways results in a scenario of transformation of progenitor and tissue stem cells into tumors through the acquirement the self-renewal property. Hematopoietic progenitor cells acquire the self-renewal property that is normally restricted to hematopoietic stem cells as a result of Dysregulation of Wnt signaling. This leads to the chronic myelogenous leukemia (CML) blast crisis [44, 45]. Further evidences revealed that most molecular and clinically distinct breast cancers originate from mammary stem cells or early progenitor cells [46, 47].

The second major concept in cancer stem cell hypothesis described tumor as containing or may be derived from specific cellular components that have the self-renewal and differentiation characteristics of stem cells. Whereas the self-renewal property is responsible for tumorigenesis, the tumor phenotypic variability is attributed to differentiation [albeit aberrant in tumors]. Recent studies and evidences supported that theory and showed the sharing of cell markers between both normal and cancer stem cells in various types of tumors. These markers include CD44, a6 integrin, h1 integrin, and CD133 (Prominin). Hematopoietic cells were extracted from patients afflicted with acute myeloid leukemia (AML) and transplanted into immune deficient mice with the intention of modeling leukemic disease *in vivo*. Results demonstrated that a core group of cells expressing CD34, a surface marker primarily associated with hematopoietic stem cells, were responsible for generation of leukemia in xenografts [8, 9, 48]. The discovery of cancer cells expressing stem cell properties also raised the possibility that cancers other than AML may be linked to CSCs [9, 49, 50].

Also, CSCs were hypothesized to exist within solid tumors as well, particularly in tissues subject to rapid self- renewal and proliferation [48]. This theory gained support when Al-Hajj and colleagues [2003] isolated putative CSCs from breast cancer, identifying a small population of CD44+CD24- cells, typical cell surface markers of normal mammary stem cells. When these CSCs were re-injected into immunodeficient mice, breast cancer arose in eight out of nine cases [10]. These cells recently showed a high angiogenic potential as they secrete a vascular endothelia growth factor (VEGF). In contrast, significantly higher numbers of other cell types derived from the tumor mass did not form tumors after injection [10, 51]. The presence of cancer stem cell in human brain tumors has elucidated and proved to have similar properties as their normal counterparts. They express the neural stem cell marker CD133 and nestin and are able of forming neurospheres *in vitro*. When injected intracranially in nonobese combined immunodeficient NOD/SCID mice, only as few as 100 of these cells could transfer tumors with the same phenotypic heterogeneity of the initial tumor [52, 53].

Reports on multiple myeloma revealed expression of syndecan-1 CD138. Only a subpopulation of cells with CD138- phenotype was clonogenic both *in vitro* and in NOD/SCID mice [54]. Similarly, subpopulation of cells in human prostate cancer with CD44+/a2h1 hi/D133+ phenotype have the stem cell properties and represent 0.1% of total tumor cells. They have the capability to form tumors in NOD/SCID mice [55, 56]. While searching for more evidences about the role of cancer stem cells in tumorigenesis, bronchial alveolar stem cells that bear the properties and multilineage differentiation properties are found at the bronchial alveolar duct junction. They thought to be responsible for the formation of lung cancers that can form tumor in mice after transforming by k-ras *in vitro* [57]. Also the role of cancer stem cells in initiation of other types of tumors such as colon cancer is elucidated where they serve as the progenitors of colorectal cancer [58, 59].

A NEW THERAPEUTIC APPROACH

Current cancer therapies are often limited by their lack of specificity and high incidence of adverse effects. Traditional chemotherapies usually rely on toxicity to cancerous and noncancerous cells as radiation therapy which cannot target specific tumor cells. The CSC theory underlies the new strategy of novel therapeutics aiming to overcome many of the limitations of current treatments, by designing therapies for the tumor-initiating cellular compartment, the CSC. Cancer and non-cancerous stem cells, express protective drug transport mechanisms that deplete the cytotoxic chemicals from the cell [60, 61]. Additionally, CSCs demonstrate enhanced radiation resistance through the increase in either the activation of DNA repair mechanisms or defense against reactive oxygen species [62, 63].

The potential therapy for CSC is to target unique cell surface markers that are expressed in stem cells, CD44, a cell surface marker [64]. They used a specific antibody to CD44 to target the cell marker and induced two effects: tumor cells were unable to effectively engraft, leading to inhibition of tumor growth, along with loss of primitive cell types within tumor grafts [65]. This study appears promising; however, a similar study indicated that treatment with antibodies had little effect on tumor formation in xenograft [66]. However, both studies were able to show that blockade of CD44 limited AML CSC proliferation, suggesting that CD44 is a required element of the AML CSC niche [64]. Another method for targeting the CSC niche involves blocking the factors that CSCs use to produce and stabilize supporting tissues. Up-regulation of VEGF has been observed in tumors, causing angiogenesis that provides the tumor with necessary blood supply [67, 68].

As with targeting the signaling pathways used by CSCs, one of the major risks to all of these therapeutic approaches is the similarities between the CSCs and other adult stem cells in treated patients [69]. Because the targeted markers and factors are present in normal adult stem cells and CSCs, these methods may be inadequately specific and have associated toxicity to normal tissue. Identification and validation of specific CSC markers through genetic, proteomic, or metabolomic screens may enable substantial progress in the development of targeted therapeutics, and antibody-mediated therapy or CSC-specific vaccine-based therapies. Therefore, study of the shared signaling pathways of CSCs may be more efficient for drug targeting opportunities [70]. It is important to note that proof-of-principle evidence in animal models and anecdotal clinical evidence exists for killing CSCs to eradicate cancer. Unique elimination of the CSC population through conditional knockdown or drug treatment in mouse models of

CML eradicated the CML, whereas drug treatments not eliminating the CSC population had little effect [71, 72]. Short-term treatment of glioblastoma multiform (GBM, brain cancer) CSCs with Notch or Hedgehog inhibitors or treatment with BMPs depletes the CSC population and prevents engraftment in immunodeficient mouse models [73, 74]. In breast cancer, the HER2-expressing cells have CSC properties, and blockade with the pharmaceutical trastuzumab depletes these CSCs preventing primary and serial engraftment into mice which explain the results in some patients [75, 76].

Characterization of adaptive and/or resistance mechanisms of CSCs to current treatment regimens is important for developing more effective future therapies. Although molecular targeted therapies show early promise in treating cancer and its stem cells [77, 78], CSCs are likely a dynamic and changing population through clonal evolution or adaptive/resistance responses. Blockade of specific CSC resistance mechanisms to traditional therapies is also a useful approach, with early successes of IL-4 blockade to sensitize colon CSCs to chemotherapy [84] and CHK kinase inhibition to radiosensitize GBM CSCs [62]. Different cancer CSCs may exhibit varied response mechanisms, such as breast *vs.* GBM CSCs in radioresistance [62, 63].

Lastly, the heterogeneity of cancer and rapid improvement in diagnostic methods is making possible an age of customized medicine for individual patients. The power of genetic and other global arrays has demonstrated repeatedly the molecular variation among cancers, and genetic signatures can often predict outcome and treatment response better than histological analysis alone [79, 80]. Molecular variation is also quickly being demonstrated in the CSCs within cancer as well [81 - 83]. Ideally, CSCs will be rapidly isolated and screened for molecular defects sensitizing them to specific drugs or subjected to a direct drug screen. Some research has focused on developing such methods [84, 85]; however, much work must be done to optimize these methods and establish feasible protocols for clinical application as the time and effort required currently make them impossible to broadly apply.

Isolation techniques of CSCs must also be further perfected before such protocols can be enacted in the clinic. With these diagnostic tools in hand, doctors may eventually be able to create a tailored prescription for oncogenic pathway components from a wide array of options, achieving individualized, targeted therapies with enhanced efficacy to patient's specific tumor with a lower toxic side effect profile. The CSC field is a relatively new one with a great deal of research and exploration still necessary to yield greater practical benefits towards improved cancer treatment. The cancer stem cell hypothesis offers a new perspective for understanding tumorigenesis, and will likely lead to rational development of refined and novel therapies that yield tangible clinical benefits to patients.

NANOTECHNOLOGY IN CANCER STEM CELL THERAPY

Nanomedicine provides a novel achievement that may cross over resistance of targeted CSCs to chemotherapy and to enhance the effects of CSCs eliminating pharmaceuticals Nanomedicine offers the loading of chemotherapeutics or active molecules on nanoparticles that act as vehicles for the delivering them specifically to targeted CSCs. The development of these Nano vehicles is a multifactorial process where size, polarity, numbers and hydrophobic and hydrophilic nature of nanoparticles must be taken into consideration [63]. They enhance therapeutic strategies targeting the unique dynamics of cancer *via* various mechanisms. These mechanisms may include prolonged circulation and improved internal distribution of the used drug resulting in its accumulation in tumor in a process known as "enhanced permeability and retention" [86]. Nano vehicles have many advantages over the conventional therapies. They can be adjusted without losing activity and can as well modulate the undesirable effect of the drug *via* encapsulation of the chemotherapeutics.

OPPORTUNITY FOR NOVEL THERAPEUTIC REGIMENS

The necessity of altering the current regimen of therapy where getting rid of cancer requires the specific targeting and elimination of CSCs. Feasibility of targeting specific pathways including those involved in self-renewal is elucidated in recent studies. For example, inhibiting g-secretase enzyme needed for Notch signaling has a great effect against breast cancer with Notch1 over expression [87, 88]. Cyclopamine and HhAntag are inhibitors of Hedgehog pathway. They show antineoplastic activity that caused a dramatic regression of tumors with no recurrence after stopping of treatment. They are nontoxic as well upon administration at least over short periods [37, 89]. Maintenance of prostate tumor-initiating cell depends on PI3K/AKT signaling. Blockage of this signaling route with NVP-BEZ235 inhibitor eliminated CD133+/CD44+tumor progenitor population *in vitro* and in xenograft tumors and sensitized prostate tumors to Taxoterein mouse models [90]. Recently, it was demonstrated that treatment with NVP-BEZ235 can inhibit EMT and radiosensitize prostate cancer cells under both hypoxic and normoxic conditions [91, 92]. NVP-BEZ235 inhibitor is currently in clinical trial for castration-resistant prostate cancer (NCT01634061, NCT01717898).

Another challenging task for development of biologically refined therapy is identification of specific biomarkers of CSCs for analysis of CSC population in pre-treatment biopsies and during the course of tumor treatment. These biomarkers of CSC are important for tumor diagnostics, for patient prognosis, prediction of patients' response to therapies and for treatments election. On the other hand, CSCs in different tumors have metabolic features, such as low mitochondrial respiration and high glycolytic activity, which may help to distinguish CSCs from their normal counter- parts [93]. Indeed, they suggested that targeting PGK1 enzyme in gastric cancer cells induce tumor stem cells differentiation. Other differentiation therapies such as bone morphogenetic proteins (BMPs) for glioblastoma and ATRA for leukemia and some solid tumors also could offer an opportunity for combination treatment modalities [94 - 97]. In addition, modification of the tumor microenvironment, including reoxygenation of CSC niche, reactivation of cytotoxic T-cells by increasing antigen cross-presentation or targeting tumor associated macrophages and fibroblasts can be another promising strategy for eradication of CSCs [98, 99].

CONCLUSION

The treatment strategies based on combination of conventional therapies targeting bulk tumor cells and CSC specific pathway inhibition bear a promise to improve cancer cure compared to monotherapies [100, 101, 102]. Taken together, current findings suggest that CSCs can potentially contribute to tumorchemo- and radioresistance and better understanding of the molecular mechanisms underlying CSC pathology and CSC interaction with their niche could be beneficial in improving cancer treatment. In many cases, tumor cell markers acquire the characteristics of their normal counterparts. Thus, differentiation between markers serving in tumor cell identification and those playing important role in stem cell behavior is important.

CONFLICT OF INTEREST

The author confirms that author has no conflict of interest to declare for this publication.

ACKNOWLEDGEMENTS

We would like to express our greatest gratitude and appreciation for Prof. Ahmed El-Hashash for affording that great chance to be a part of this great scientific product.

REFERENCES

[1] Sell S. Stem cell origin of cancer and differentiation therapy. Crit Rev Oncol Hematol 2004; 51(1): 1-28.
[PMID: 15207251]

[2] Tubiana M, Malaise EP. Cell proliferation kinetics of tumors and cancer treatment. Pathol Biol (Paris) 1973; 21(6): 647-64.
[PMID: 4581197]

[3] Fearon ER, Vogelstein B. A genetic model for colorectal tumorigenesis. Cell 1990; 61(5): 759-67.
[PMID: 2188735]

[4] Hanahan D, Weinberg RA. The hallmarks of cancer. Cell 2000; 100(1): 57-70.
[PMID: 10647931]

[5] Reya T, Morrison SJ, Clarke MF, Weissman IL. Stem cells, cancer, and cancer stem cells. Nature 2001; 414(6859): 105-11.
[PMID: 11689955]

[6] Oberling C. The riddle of cancer. New Haven: Yale University Press 1952; p. 238.

[7] Rather L. The genesis of cancer. Baltimore: Johns Hopkins University Press 1978; p. 262.

[8] Lapidot T, Sirard C, Vormoor J, *et al.* A cell initiating human acute myeloid leukaemia after transplantation into SCID mice. Nature 1994; 367(6464): 645-8.
[PMID: 7509044]

[9] Bonnet D, Dick JE. Human acute myeloid leukemia is organized as a hierarchy that originates from a

primitive hematopoietic cell. Nat Med 1997; 3(7): 730-7.
[PMID: 9212098]

[10]   Al-Hajj M, Wicha MS, Benito-Hernandez A, Morrison SJ, Clarke MF. Prospective identification of tumorigenic breast cancer cells. Proc Natl Acad Sci USA 2003; 100(7): 3983-8.
[PMID: 12629218]

[11]   Rich JN, Bao S. Chemotherapy and cancer stem cells. Cell Stem Cell 2007; 1(4): 353-5.
[PMID: 18371369]

[12]   Al-Hajj M, Becker MW, Wicha M, Weissman I, Clarke MF. Therapeutic implications of cancer stem cells. Curr Opin Genet Dev 2004; 14(1): 43-7.
[PMID: 15108804]

[13]   Nuciforo P, Fraggetta F. Cancer stem cell theory: pathologists' considerations and ruminations about wasting time and wrong evaluations. J Clin Pathol 2004; 57(7): 782-95.
[PMID: 15220380]

[14]   Hanahan D, Weinberg RA. Hallmarks of cancer: the next generation. Cell 2011; 144(5): 646-74.
[PMID: 21376230]

[15]   Siegel R, Naishadham D, Jemal A. Cancer statistics, 2013. CA Cancer J Clin 2013; 63(1): 11-30.
[PMID: 23335087]

[16]   Hwang-Verslues WW, Kuo W-H, Chang P-H, *et al.* Multiple lineages of human breast cancer stem/progenitor cells identified by profiling with stem cell markers. PLoS One 2009; 4(12): e8377.
[PMID: 20027313]

[17]   Navin N, Hicks J. Future medical applications of single-cell sequencing in cancer. Genome Med 2011; 3(5): 31-7.
[PMID: 21631906]

[18]   Notta F, Mullighan CG, Wang JC, *et al.* Evolution of human BCR-ABL1 lymphoblastic leukaemia-initiating cells. Nature 2011; 469(7330): 362-7.
[PMID: 21248843]

[19]   Gerlinger M, Rowan AJ, Horswell S, *et al.* Intratumor heterogeneity and branched evolution revealed by multiregion sequencing. N Engl J Med 2012; 366(10): 883-92.
[PMID: 22397650]

[20]   Beà S, Valdés-Mas R, Navarro A, *et al.* Landscape of somatic mutations and clonal evolution in mantle cell lymphoma. Proc Natl Acad Sci USA 2013; 110(45): 18250-5.
[PMID: 24145436]

[21]   Sottoriva A, Spiteri I, Piccirillo SG, *et al.* Intratumor heterogeneity in human glioblastoma reflects cancer evolutionary dynamics. Proc Natl Acad Sci USA 2013; 110(10): 4009-14.
[PMID: 23412337]

[22]   Zhu Y, Ghosh P, Charnay P, Burns DK, Parada LF. Neurofibromas in NF1: Schwann cell origin and role of tumor environment. Science 2002; 296(5569): 920-2.
[PMID: 11988578]

[23]   Wicha MS, Liu S, Dontu G. Cancer stem cells: an old idea--a paradigm shift. Cancer Res 2006; 66(4): 1883-90.

[PMID: 16488983]

[24]   Li L, Neaves WB. Normal stem cells and cancer stem cells: the niche matters. Cancer Res 2006; 66(9): 4553-7.
        [PMID: 16651403]

[25]   Patel SA, Heinrich AC, Reddy BY, Rameshwar P. Inflammatory mediators: Parallels between cancer biology and stem cell therapy. J Inflamm Res 2009; 2: 13-9.
        [PMID: 20428325]

[26]   Ishikawa F, Yoshida S, Saito Y, *et al.* Chemotherapy-resistant human AML stem cells home to and engraft within the bone-marrow endosteal region. Nat Biotechnol 2007; 25(11): 1315-21.
        [PMID: 17952057]

[27]   Cho K, Wang X, Nie S, Chen ZG, Shin DM. Therapeutic nanoparticles for drug delivery in cancer. Clin Cancer Res 2008; 14(5): 1310-6.
        [PMID: 18316549]

[28]   Quintana E, Shackleton M, Sabel MS, Fullen DR, Johnson TM, Morrison SJ. Efficient tumour formation by single human melanoma cells. Nature 2008; 456(7222): 593-8.
        [PMID: 19052619]

[29]   Tang DG. Understanding cancer stem cell heterogeneity and plasticity. Cell Res 2012; 22(3): 457-72.
        [PMID: 22357481]

[30]   Gupta PB, Fillmore CM, Jiang G, *et al.* Stochastic state transitions give rise to phenotypic equilibrium in populations of cancer cells. Cell 2011; 146(4): 633-44.
        [PMID: 21854987]

[31]   Cojoc M, Mäbert K, Muders MH, Dubrovska A. A role for cancer stem cells in therapy resistance: cellular and molecular mechanisms. Semin Cancer Biol 2015; 31: 16-27.
        [PMID: 24956577]

[32]   Little MP, Boice JD Jr. Comparison of breast cancer incidence in the Massachusetts tuberculosis fluoroscopy cohort and in the Japanese atomic bomb survivors. Radiat Res 1999; 151(2): 218-24.
        [PMID: 9952307]

[33]   Smith GH, Chepko G. Mammary epithelial stem cells. Microsc Res Tech 2001; 52(2): 190-203.
        [PMID: 11169867]

[34]   Liu S, Dontu G, Wicha MS. Mammary stem cells, self-renewal pathways, and carcinogenesis. Breast Cancer Res 2005; 7(3): 86-95.
        [PMID: 15987436]

[35]   Dontu G, Jackson KW, McNicholas E, Kawamura MJ, Abdallah WM, Wicha MS. Role of Notch signaling in cell-fate determination of human mammary stem/progenitor cells. Breast Cancer Res 2004; 6(6): R605-15.
        [PMID: 15535842]

[36]   Unden AB, Holmberg E, Lundh-Rozell B, *et al.* Mutations in the human homologue of Drosophila patched (PTCH) in basal cell carcinomas and the Gorlin syndrome: different *in vivo* mechanisms of PTCH inactivation. Cancer Res 1996; 56(20): 4562-5.
        [PMID: 8840960]

[37]   Karhadkar SS, Bova GS, Abdallah N, *et al.* Hedgehog signalling in prostate regeneration, neoplasia and metastasis. Nature 2004; 431(7009): 707-12.
[PMID: 15361885]

[38]   Olsen CL, Hsu PP, Glienke J, Rubanyi GM, Brooks AR. Hedgehog-interacting protein is highly expressed in endothelial cells but down-regulated during angiogenesis and in several human tumors. BMC Cancer 2004; 4: 43-9. [electronic resource].
[PMID: 15294024]

[39]   Diévart A, Beaulieu N, Jolicoeur P. Involvement of Notch1 in the development of mouse mammary tumors. Oncogene 1999; 18(44): 5973-81.
[PMID: 10557086]

[40]   Siziopikou K, Miao H, Rizzo P, *et al.* Notch signaling is a therapeutic target in breast cancer. Proceedings of the 94th Annual Meeting of the AACR. 1277-8.

[41]   Nam Y, Aster JC, Blacklow SC. Notch signaling as a therapeutic target. Curr Opin Chem Biol 2002; 6(4): 501-9.
[PMID: 12133727]

[42]   Nickoloff BJ, Osborne BA, Miele L. Notch signaling as a therapeutic target in cancer: a new approach to the development of cell fate modifying agents. Oncogene 2003; 22(42): 6598-608.
[PMID: 14528285]

[43]   Benson RA, Lowrey JA, Lamb JR, Howie SE. The Notch and Sonic hedgehog signalling pathways in immunity. Mol Immunol 2004; 41(6-7): 715-25.
[PMID: 15220006]

[44]   Jamieson CH, Ailles LE, Dylla SJ, *et al.* Granulocyte-macrophage progenitors as candidate leukemic stem cells in blast-crisis CML. N Engl J Med 2004; 351(7): 657-67.
[PMID: 15306667]

[45]   Kelly LM, Gilliland DG. Genetics of myeloid leukemias. Annu Rev Genomics Hum Genet 2002; 3: 179-98.
[PMID: 12194988]

[46]   Dontu G, Al-Hajj M, Abdallah WM, Clarke MF, Wicha MS. Stem cells in normal breast development and breast cancer. Cell Prolif 2003; 36 (Suppl. 1): 59-72.
[PMID: 14521516]

[47]   Li Y, Welm B, Podsypanina K, *et al.* Evidence that transgenes encoding components of the Wnt signaling pathway preferentially induce mammary cancers from progenitor cells. Proc Natl Acad Sci USA 2003; 100(26): 15853-8.
[PMID: 14668450]

[48]   Dalerba P, Cho RW, Clarke MF. Cancer stem cells: models and concepts. Annu Rev Med 2007; 58: 267-84.
[PMID: 17002552]

[49]   Hope KJ, Jin L, Dick JE. Acute myeloid leukemia originates from a hierarchy of leukemic stem cell classes that differ in self-renewal capacity. Nat Immunol 2004; 5(7): 738-43.
[PMID: 15170211]

[50]   Blagosklonny MV. Target for cancer therapy: proliferating cells or stem cells. Leukemia 2006; 20(3): 385-91.
[PMID: 16357832]

[51]   Ponti D, Costa A, Zaffaroni N, *et al.* Isolation and *in vitro* propagation of tumorigenic breast cancer cells with stem/progenitor cell properties. Cancer Res 2005; 65(13): 5506-11.
[PMID: 15994920]

[52]   Singh SK, Hawkins C, Clarke ID, *et al.* Identification of human brain tumour initiating cells. Nature 2004; 432(7015): 396-401.
[PMID: 15549107]

[53]   Singh SK, Clarke ID, Hide T, Dirks PB. Cancer stem cells in nervous system tumors. Oncogene 2004; 23(43): 7267-73.
[PMID: 15378086]

[54]   Matsui W, Huff CA, Wang Q, *et al.* Characterization of clonogenic multiple myeloma cells. Blood 2004; 103(6): 2332-6.
[PMID: 14630803]

[55]   Richardson GD, Robson CN, Lang SH, Neal DE, Maitland NJ, Collins AT. CD133, a novel marker for human prostatic epithelial stem cells. J Cell Sci 2004; 117(Pt 16): 3539-45.
[PMID: 15226377]

[56]   Maitland NJ, Collins AT, Bryce S, *et al.* Prospective identification of tumorigenic prostate cancer stem cells. Proceedings of the 96th Annual Meeting of the AACR. 10946.

[57]   Kim CF, Jackson EL, Woolfenden AE, *et al.* Identification of bronchioalveolar stem cells in normal lung and lung cancer. Cell 2005; 121(6): 823-35.
[PMID: 15960971]

[58]   Liu C, Zhao G, Liu J, *et al.* Novel biodegradable lipid nano complex for siRNA delivery significantly improving the chemosensitivity of human colon cancer stem cells to paclitaxel. J Control Release 2009; 140(3): 277-83.
[PMID: 19699770]

[59]   Ricci-Vitiani L, Lombardi DG, Pilozzi E, *et al.* Identification and expansion of human colon-cance--initiating cells. Nature 2007; 445(7123): 111-5.
[PMID: 17122771]

[60]   Dean M, Fojo T, Bates S. Tumour stem cells and drug resistance. Nat Rev Cancer 2005; 5(4): 275-84.
[PMID: 15803154]

[61]   Blagosklonny MV. Target for cancer therapy: proliferating cells or stem cells. Leukemia 2006; 20(3): 385-91.
[PMID: 16357832]

[62]   Bao S, Wu Q, McLendon RE, *et al.* Glioma stem cells promote radioresistance by preferential activation of the DNA damage response. Nature 2006; 444(7120): 756-60.
[PMID: 17051156]

[63]   Diehn M, Cho RW, Lobo NA, *et al.* Association of reactive oxygen species levels and radioresistance in cancer stem cells. Nature 2009; 458(7239): 780-3.

[PMID: 19194462]

[64]   Williams DA, Cancelas JA. Leukaemia: niche retreats for stem cells. Nature 2006; 444(7121): 827-8.
        [PMID: 17167463]

[65]   Jin L, Hope KJ, Zhai Q, Smadja-Joffe F, Dick JE. Targeting of CD44 eradicates human acute myeloid
        leukemic stem cells. Nat Med 2006; 12(10): 1167-74.
        [PMID: 16998484]

[66]   Krause DS, Lazarides K, von Andrian UH, Van Etten RA. Requirement for CD44 in homing and
        engraftment of BCR-ABL-expressing leukemic stem cells. Nat Med 2006; 12(10): 1175-80.
        [PMID: 16998483]

[67]   Kim KJ, Li B, Winer J, *et al.* Inhibition of vascular endothelial growth factor-induced angiogenesis
        suppresses tumour growth in vivo. Nature 1993; 362(6423): 841-4.
        [PMID: 7683111]

[68]   Plate KH, Breier G, Weich HA, Risau W. Vascular endothelial growth factor is a potential tumour
        angiogenesis factor in human gliomas in vivo. Nature 1992; 359(6398): 845-8.
        [PMID: 1279432]

[69]   Reya T, Clevers H. Wnt signalling in stem cells and cancer. Nature 2005; 434(7035): 843-50.
        [PMID: 15829953]

[70]   Zhou BB, Zhang H, Damelin M, Geles KG, Grindley JC, Dirks PB. Tumour-initiating cells:
        challenges and opportunities for anticancer drug discovery. Nat Rev Drug Discov 2009; 8(10): 806-23.
        [PMID: 19794444]

[71]   Pérez-Caro M, Cobaleda C, González-Herrero I, *et al.* Cancer induction by restriction of oncogene
        expression to the stem cell compartment. EMBO J 2009; 28(1): 8-20.
        [PMID: 19037256]

[72]   Ito K, Bernardi R, Morotti A, *et al.* PML targeting eradicates quiescent leukaemia-initiating cells.
        Nature 2008; 453(7198): 1072-8.
        [PMID: 18469801]

[73]   Bar EE, Chaudhry A, Lin A, *et al.* Cyclopamine-mediated hedgehog pathway inhibition depletes stem-
        like cancer cells in glioblastoma. Stem Cells 2007; 25(10): 2524-33.
        [PMID: 17628016]

[74]   Fan X, Khaki L, Zhu TS, *et al.* NOTCH pathway blockade depletes CD133-positive glioblastoma cells
        and inhibits growth of tumor neurospheres and xenografts. Stem Cells 2010; 28(1): 5-16.
        [PMID: 19904829]

[75]   Korkaya H, Wicha MS. HER-2, notch, and breast cancer stem cells: targeting an axis of evil. Clin
        Cancer Res 2009; 15(6): 1845-7.
        [PMID: 19276254]

[76]   Magnifico A, Albano L, Campaner S, *et al.* Tumor-initiating cells of HER2-positive carcinoma cell
        lines express the highest oncoprotein levels and are sensitive to trastuzumab. Clin Cancer Res 2009;
        15(6): 2010-21.
        [PMID: 19276287]

[77]   Soeda A, Inagaki A, Oka N, *et al.* Epidermal growth factor plays a crucial role in mitogenic regulation of human brain tumor stem cells. J Biol Chem 2008; 283(16): 10958-66.
[PMID: 18292095]

[78]   Hambardzumyan D, Becher OJ, Rosenblum MK, Pandolfi PP, Manova-Todorova K, Holland EC. PI3K pathway regulates survival of cancer stem cells residing in the perivascular niche following radiation in medulloblastoma *in vivo*. Genes Dev 2008; 22(4): 436-48.
[PMID: 18281460]

[79]   Phillips HS, Kharbanda S, Chen R, *et al.* Molecular subclasses of high-grade glioma predict prognosis, delineate a pattern of disease progression, and resemble stages in neurogenesis. Cancer Cell 2006; 9(3): 157-73.
[PMID: 16530701]

[80]   Liu R, Wang X, Chen GY, *et al.* The prognostic role of a gene signature from tumorigenic breast-cancer cells. N Engl J Med 2007; 356(3): 217-26.
[PMID: 17229949]

[81]   Gunther HS, Schmidt NO, Phillips HS, Kemming D, Kharbanda S, Soriano R, *et al.* Glioblastoma-derived stem cell-enriched cultures form distinct subgroups according to molecular and phenotypic criteria. Oncogene 2007; 26: 122-9.
[PMID: 18037961]

[82]   Okamoto OK, Carvalho AC, Marti LC, Vêncio RZ, Moreira-Filho CA. Common molecular pathways involved in human CD133+/CD34+ progenitor cell expansion and cancer. Cancer Cell Int 2007; 7: 11-9.
[PMID: 17559657]

[83]   Hwang-Verslues WW, Kuo WH, Chang PH, *et al.* Multiple lineages of human breast cancer stem/progenitor cells identified by profiling with stem cell markers. PLoS One 2009; 4(12): e8377.
[PMID: 20027313]

[84]   Gal H, Makovitzki A, Amariglio N, Rechavi G, Ram Z, Givol D. A rapid assay for drug sensitivity of glioblastoma stem cells. Biochem Biophys Res Commun 2007; 358(3): 908-13.
[PMID: 17512905]

[85]   Pollard SM, Yoshikawa K, Clarke ID, *et al.* Glioma stem cell lines expanded in adherent culture have tumor-specific phenotypes and are suitable for chemical and genetic screens. Cell Stem Cell 2009; 4(6): 568-80.
[PMID: 19497285]

[86]   Woodward WA, Chen MS, Behbod F, Alfaro MP, Buchholz TA, Rosen JM. WNT/beta-catenin mediates radiation resistance of mouse mammary progenitor cells. Proc Natl Acad Sci USA 2007; 104(2): 618-23.
[PMID: 17202265]

[87]   Weijzen S, Rizzo P, Braid M, *et al.* Activation of Notch-1 signaling maintains the neoplastic phenotype in human Ras-transformed cells. Nat Med 2002; 8(9): 979-86.
[PMID: 12185362]

[88]   Pece S, Serresi M, Santolini E, *et al.* Loss of negative regulation by Numb over Notch is relevant to

human breast carcinogenesis. J Cell Biol 2004; 167(2): 215-21.
[PMID: 15492044]

[89]   Romer JT, Kimura H, Magdaleno S, *et al.* Suppression of the Shh pathway using a small molecule inhibitor eliminates medulloblastoma in Ptc1(+/-)p53(-/-) mice. Cancer Cell 2004; 6(3): 229-40.
[PMID: 15380514]

[90]   Dubrovska A, Kim S, Salamone RJ, *et al.* The role of PTEN/Akt/PI3K signaling in the maintenance and viability of prostate cancer stem-like cell populations. Proc Natl Acad Sci USA 2009; 106(1): 268-73.
[PMID: 19116269]

[91]   Chang L, Graham PH, Hao J, *et al.* Acquisition of epithelial-mesenchymal transition and cancer stem cell phenotypes is associated with activation of the PI3K/Akt/mTOR pathway in prostate cancer radioresistance. Cell Death Dis 2013; 4: e875.
[PMID: 24157869]

[92]   Potiron VA, Abderrahmani R, Giang E, *et al.* Radiosensitization of prostate cancer cells by the dual PI3K/mTOR inhibitor BEZ235 under normoxic and hypoxic conditions. Radiother Oncol 2013; 106(1): 138-46.
[PMID: 23321494]

[93]   Pecqueur C, Oliver L, Oizel K, Lalier L, Vallette FM. Targeting metabolism to induce cell death in cancer cells and cancer stem cells. Int J Cell Biol 2013.

[94]   Formelli F, Cleris L. Synthetic retinoid fenretinide is effective against a human ovarian carcinoma xenograft and potentiates cisplatin activity. Cancer Res 1993; 53(22): 5374-6.
[PMID: 8221674]

[95]   Fenaux P, Chastang C, Chevret S, *et al.* A randomized comparison of all transretinoic acid (ATRA) followed by chemotherapy and ATRA plus chemotherapy and the role of maintenance therapy in newly diagnosed acute promyelocytic leukemia. Blood 1999; 94(4): 1192-200.
[PMID: 10438706]

[96]   Piccirillo SG, Reynolds BA, Zanetti N, *et al.* Bone morphogenetic proteins inhibit the tumorigenic potential of human brain tumour-initiating cells. Nature 2006; 444(7120): 761-5.
[PMID: 17151667]

[97]   Zieker D, Bühler S, Ustündag Z, *et al.* Induction of tumor stem cell differentiation--novel strategy to overcome therapy resistance in gastric cancer. Langenbecks Arch Surg 2013; 398(4): 603-8.
[http://dx.doi.org/10.1007/s00423-013-1058-5] [PMID: 23412594]

[98]   Reisfeld RA. The tumor microenvironment: a target for combination therapy of breast cancer. Crit Rev Oncog 2013; 18(1-2): 115-33.
[PMID: 23237555]

[99]   Yang X, Zhang X, Fu ML, *et al.* Targeting the tumor microenvironment with interferon-β bridges innate and adaptive immune responses. Cancer Cell 2014; 25(1): 37-48.
[PMID: 24434209]

[100]   Baumann M, Krause M, Hill R. Exploring the role of cancer stem cells in radioresistance. Nat Rev Cancer 2008; 8(7): 545-54.

[PMID: 18511937]

[101] Linkous AG, Yazlovitskaya EM. Novel radiosensitizing anticancer therapeutics. Anticancer Res 2012; 32(7): 2487-99.
[PMID: 22753705]

[102] Shigdar S, Lin J, Li Y, *et al.* Cancer stem cell targeting: the next generation of cancer therapy and molecular imaging. Ther Deliv 2012; 3(2): 227-44.
[PMID: 22834199]

CHAPTER 9

Stem Cells and Neurodegenerative Diseases

Hanaa H. Ahmed*, Hadeer A. Aglan and **Ahmed A. Abd-Rabou**

Hormones Department, Medical Research Division, National Research Center, Cairo, Egypt

Abstract: Neurodegenerative diseases are a major concern of our present time and which underpin the ageing era that invades the world. The Neurodegenerative diseases are caused by certain neuronal loss in specific regions of the brain. Alzheimer's disease (AD) and Parkinson's disease (PD) are some of the examples of the neurodegenerative diseases that have no fundamental cure available. Some drug treatments can alleviate the symptoms associated with the neurodegenerative diseases. However, they do not tackle the main pathological factors and cannot be clinically suitable for all patients. Moreover, they are not affordable by all patients as a long term medicine. Therefore, developing new and effective medications for AD and PD is deemed necessary. Recent research has aimed at developing therapies that modify the disease. These therapies perform their actions by interacting with the pathophysiologic cascade in order to postpone the disease onset or prevents the progression from occurring on a fast pace. Embryonic and Adult stem cells have demonstrated high therapeutic potential for tissue regeneration. As well, cell replacement therapy would introduce cure for these neurological disorders. Treatment of neurodegenerative diseases using stem cell transplantation has attracted a great deal of attention lately. This is owing to the fact that stem cells are readily available, can be easily expanded in culture, and can have sustainability when transplanted for relatively long periods of time. Moreover, the growth factors and cytokines released by stem cells facilitate neo-vascularization of damaged tissue leading to neurogenesis, as well as affording anti-inflammatory, anti-apoptotic, and anti-oxidative effects among other reparative responses. From this point of view, stem cell therapy will provide a powerful and effective cure for most of neurodeteriorative diseases in the near future.

* **Correspondence author Hanaa H. Ahmed:** Medical Research Division, National Research Center, 12622, Cairo, Egypt; Tel: +2 01223661935; Fax: 02-33370931; E-mail: hanaaomr@yahoo.com

Ahmed El-Hashash (Ed)

Keywords: Alzheimer's disease, Neurodegenerative diseases, Parkinson's disease, Stem cells.

1. INTRODUCTION

Regenerative medicine is an evolving therapeutic research era with several functions as to restore, maintain and improve body functions [1]. The regenerative medicine can help repair cells, tissues and organs in order to perform their functions successfully [2]. Stem cells present naturally in every tissue type function mainly as a regenerating and repairing system. Hence, the future of the translational medicine relies mainly on stem cells as a promising tool in this regard [3]. Neurodegeneration is a scientific term used to describe the progressive loss of structure or function of neurons, which includes the neural death [4].

Neurodegenerative diseases that are exemplified in Alzheimer's, Parkinson's, Huntington's diseases, and amyotrophic lateral sclerosis comprise a group of pathologies that possess a separated etiology with specific morphological and pathophysiological features. The main neuropathological features that underlie these disorders are represented in abnormal protein dynamics associated with defective protein degradation and aggregation, oxidative stress associating with formation of free radicals, mitochondrial dysfunctions, and neuroinflammatory processes [5]. These neuro- associated-diseases are characterized by a progressive and specific loss of certain brain cells [6]. For example, loss of a group of neurons called the "nucleus basalis" in the "substantia innominata" of the basal forebrain with enrichment in acetylcholine (Ach) and choline acetyl-transferase (ChAT) is linked to Alzheimer's disease (AD). While, in Parkinson's disease (PD), neurons die in the "substantia nigra", a brain structure located in the midbrain, that plays an important role in reward, addiction and movement.

The mechanisms that underlie the neuronal cell death in the nucleus basalis in AD and in the substantia nigra in PD are not yet clear. As a matter of fact, many factors represented in alterations in the energy modes of degenerating neurons, the presence of abnormal aggregated proteins (ß-amyloid and tau proteins in AD and a-synuclein and parkin in PD), abnormalities in the function of the ubiquitin-proteasome system, a lack of trophic factors, instabilities in the levels of cytokines, coupled with the ionic gradient disruption and the signal transduction processes, may participate in the specificity of neurodegenerative processes [7, 8]. European scientists found that the various types of blood cells are derived from a specific stem cell. Stem cells were first investigated by Becker *et al.* [9], who used irradiated mice in his study and injected them with bone marrow cells and concluded that each nodule derived from a single marrow cell is capable of infinite self renewal. Therefore, the self regeneration ability and the ability to differentiate into different cell lineages (potency) are two essential traits of stem cells. Right signals triggerstem cells differentiation to many different sorts of cells that form the organism [10]. Potency specifies the differential potential of the stem cells. Totipotent stem cells are a result of the fusion of an egg and a sperm cell. The first few divisions of the fertilized egg give rise to totipotent stem cells, as well. These cells can differentiate into different cell types such as embryonic and extraembryonic cells.

Pleuripotent stem cells are the descendents of totipotent cells and they have the ability to differentiate into cells derived from three germ layers. Pleuripotent stem cells arise as an internal mass of cells within a blastocyst (Blastula). Blastocyst is a sphere that has a thin hollow wall and is composed of an external layer of cells, a fluid filled cavity and an inward cell mass containing pleuripotent stem cells. Multipotent stem cells have the ability to produce cells that belong to the same family of cells (*e.g.* hematopoeticstem cells can give rise to red blood cells, white blood cells, platelets, *etc*). Unipotent stem cells can give rise to a single cell type and they have the capability of renewing themselves, which varies them from non-stem cells [11]. The potential of stem cell sand their plasticity are having invaluable properties for regenerative medicine [12].

2. ROLE OF STEM CELLS IN THE REGENERATIVE MEDICINE

Embryonic and non-embryonic stem cells (ESCs and non-ESCs) are two major categories of the stem cells. ESCs are totipotent in nature and have the ability to differentiate into the three embryonic germ layers. On the other hand, non-ESCs, or adult stem cells, are just multipotent; their capability to differentiate into different kinds of cells seems to be more limited [13].

2.1. Embryonic Stem Cells

Human ESCs are self-renewable primitive cells and can be differentiated into the different types of cells that are present in adult human body [14]. The mouse ESCs were first isolated in 1981 [15, 16], but in 1998, a modern era developed in stem cell biology that revolutionized the isolation of cells from human blastocysts and fetal tissue with their specific capability of differentiating into all bodycells [17]. ESCs are produced from embryos at an early stage of development prior to the implantation period that naturally occurs in the uterus.

The primary stage of differentiation takes place in nearly five days of development, when an external layer of cells, that forms part of the placenta (trophectoderm), detaches from the inner cell mass (ICM) [10].

2.1.1. Embryonic Germ Cells

Embryonic germ cells (EGCs) are derived from primordial germ line cells in early fetal tissue. The data available for the animal embryonic germ cell experiments are limited when compared with those available for the ESC. In addition, the possible outcomes will be some what limited compared to ESCs, since the embryonic germ cells are more developed (5-9 weeks) [18]. In 1998, the traits of the germ cells derived from the gonadal ridge of human tissue were reported [19].

2.1.2. Amniotic Epithelial Cells

A diverse group of cells are displaced into the amniotic fluid that surrounds the embryo during development. The Amniotic membrane present in human placenta acts as a source of the Amniotic epithelial cells (AECs) that express the markers of pluripotent ESCs and EGCs, such as octamer-binding transcription factor 4 (Oct-4), Nanog, and alkaline phosphatase. Moreover, they can be differentiated *in vitro* as ESCs and EGCs, from three germ layers in the cell lineage (generations), including pancreatic endocrine cells and hepatocytes (endoderm), cardiomyocytes (mesoderm), and neural cells (ectoderm) [20, 21].

2.1.3. Embryonic Stem Cells and Regenerative Medicine

There is a growing interest in emerging the stem cell benefits following transplantations of neural progenitors and cardiomyocytes derived from human ESCs in some disorders as Parkinson's disease and myocardial injuryin animal models respectively [12, 22, 23]. Additionally, scientists used ESC-derived tissues for treatment of Parkinson's disease models [12, 24], myocardial infarction models [25], and spinal injury models [26]. Considering neurodegenerative disorders, it is more favorable to transplant neural progenitor cells or developed neuronal cells. In contrast, current studies indicated that even in the intact state, human ESCs express certain levels of human leukocyte antigen (HLA) class I antigens that elevate as the mature cells [14, 27]. Intriguingly, some studies investigated that the cell division irreversible termination is associated with organism death of the embryo [28, 29]. ESCs are pluripotent cells that can generate products of all three embryonic germ layers. Owing to their unique properties, ESCs that are specific to patients are vital for transplantation therapies, taking in consideration the prospective clinical use of the regenerative medicine [30].

2.2. Adult Stem Cells

Adult stem cells, or non-ESCs, are located in the developed tissue of organisms belonging to any age group. Non-ESCs comprise different populations of undifferentiated cells and are present in the majority of adult tissues, where they act as compartments during the regular turnover of an organ or tissue. Both their potency and proliferative capability are usually more limited than those of ESCs. Non-ESCs are embedded within body organs and are surrounded by millions of normal cells. They have the ability to renew damaged or dead body cells when necessary. Recently, it had been thought that a hematopoietic stem cell can not divide to give a very different tissue, such as brain neurons. Therefore, it is deemed necessary to research using of non-ECSs for the different cell-based therapeutic strategies [10]. The definition of adult stem cells is governed by several factors such as their ability to self-renew and their capability to proliferate and differentiate into a minimum of one kind of mature functional progeny [31].

2.2.1. Hematopoietic Stem Cells

Hematopoietic stem cells (HSCs) possess specific morphologic characteristics and cell- surface markers [32] that enable them to be tracked in the blood circulation and isolated to be used in cell culture *in vitro* [18]. HSCs are present in the bone marrow by a percentage of less than 0.05% of the total mass. However, they are capable of regenerating the lineages that form the blood. This specific property results in a wide scale of clinical advantages, promoting the HSC compartment as the best characterized stem cell niche [33].

2.2.2. Neural Stem Cells

Neural stem cells (NSCs) are constantly self-renewable and are blessed with the power of generating intermediate and mature cells of both glial and neuronal lineages [34]. Generally, the adults and embryos brains can produce NSCs. Adult neural stem cells have the capability to differentiate into numerous sorts of brain cells, such as oligodendrocytes, astrocytes, and neurons [10]. In addition, these NSCs may be described as multipotent as they could participate in producing several tissue types in embryos, when injected into mice blastocysts of mice [35]. NSCs, in brain-injury models, proliferate in those neurogenic regions and can migrate toward the damaged site [36].

2.2.3. Mesenchymal Stem Cells

Mesenchymal stem cells (MSCs) are stromal stem cells that are heterogeneous in nature; they are multipotent and possess a self-renewal property. In addition, MSCs could differentiate into cells of the mesodermal lineages and other embryonic lineages, including osteocytes, chondrocytes, adipocytes, muscle cells, hepatocytes, nerve cells, and epithelial cells [37].

It has been confirmed by research that MSCs have potential immunomodulation [38], anti-inflammation characteristics [39]. Additionally, they contribute to producing wide range of cytokines and growth factors *in vitro* and *in vivo* that have been investigated to play an important role in recovering injured tissues [40], which includes functional recovery [41]. Currently, clinical trials on MSCs have highlighted the pivotal role played by MSCs in tissue regeneration and in treating different disorders [37]. MSCs have several advantages over ESCs, hematological, and neural stem cells, which can be represented in their accessibility, little immunogenicity and few ethical concerns. The MSCs autologous origin protects against the risk of immune rejection and their nature as adult cells, reduces the development of tumors [42]. Moreover, they possess the property of proliferating on a wide scale *in vitro* without losing their undifferentiated multipotent property [43].

These characteristics make MSCs perfect fit for vital therapeutic strategies that involve tissue engineering, regenerative medicine and autoimmune disease therapy. A great deal of research has clarified that the therapeutic function played by MSCs relies on their ability to differentiate to repair damaged tissue, and on their potency to modulate local environment, activate endogenous progenitor cells, and secrete various factors [44]. To date, bone marrow, adipose tissue, umbilical cord blood, and umbilical cord have usually been regarded as the main sources of MSCs for tissue regeneration and engineering [45].

2.2.3.1. Bone Marrow-Derived MSCs

Adult bone marrow is a potentially rich source of stem cells and contains two types of prototypical stem cell populations: hematopoietic stem cells and mesenchymal stem cells. Bone marrow-derived MSCs (BM-MSCs) have great therapeutic power since they can be simply separated from bone marrow and can expand with a fast pace *in vitro* for autologous transplantation.

They are allogeneic and non-immunogenic as well, thus the risk of rejection is avoided [46]. Moreover, BM-MSCs have the ability to survive for a long time (approximately 5 months) after *in vivo* transplantation [47]. Recently, very often BM-MSCs serve as a positive control for MSCs isolated from other tissues, owing to most data on MSCs derives from studies performed on BM- MSCs [48].

2.2.3.2. Adipose Tissue-Derived MSCs

Very recently, it was found that adipose tissue is a convenient source of MSCs [49]. Adipose tissue is widely available and can be obtained easily in huge amounts with minimum patient inconvenience, thus adipose tissue-derived MSCs (ADMSCs) can supply an alternative excessive source of MSCs for stem cell therapy [50]. Also, ADMSCs provide several advantages over other multipotent cells including easy extraction procedures, high rate of proliferation, and high abundance [51].

2.2.3.3. Umblical Cord Blood-Derived MSCs

Although, umblical cord blood-derived MSCs (UCB-MSCs) are particularly rare and the degree of isolating the MSC successfully is low (<30%) [37], UCB could act as another source of allogeneic MSCs because of their isolation ease [52].

2.2.3.4. Amniotic Membrane-Derived MSCs

Amniotic membrane is considered as an available source of MSCs [53]. MSCs can be taken from amniotic membrane, which is usually disposed, without any interfering procedures and holding no ethical issues. Similar to BM-MSCs, MSc derived from amniotic membrane show low immunogenicity, multi-potent differentiation capability, and anti-inflammatory ability [54].

2.2.3.5. Migration and Homing of MSCs

Various *in vivo* studies have shown that MSCs can perform their therapeutic function by migrating to injured and inflamed tissues [55]. The mechanism by which MSCs migrate effectively might include: (1) specific ligands up regulated by injury tissues, which do not only ease adhesion and trafficking of MSCs, but also provide MSCs with a microenvironment or niche that has a specific action in supporting their self-renewal and maintaining their multi-capability [56]; (2) integrins, selectins, and chemokine receptors [57] that are expressed on the surface of the MSCs, and play an important role in the flow of MSCs across the endothelium, and (3) MSCs are locked in a passive way inside micro-capillaries including arterioles and post-capillary venules, and then they interact directly with supplementary cells and produce a wide range of trophic cytokines and soluble growth factors [58, 59]. An *in vitro* study showed that murine MSCs migrate towards cells that are separated from mouse lungs that were injured by bleomycin, but not towards lungs cells of healthy mouse. Chemotaxis test showed that cultured MSCs flow towards a variety of growth factors and chemokines in a fashion that depends on their dose *in vitro*, and that pro-inflammatory cytokine tumor necrosis factor-a provoke the migration of the chemokine and this illustrates that infused MSCs can be systemically directed to inflammatory sites [60].

2.2.3.6. MSCs and Neurodegenerative Diseases

Intriguingly, it was suggested that MSCs have a pivotal curative role for acute injury and central nervous system (CNS) diseases of progressive and degenerative nature [61], stroke [62], autoimmune encephalomyelitis (EAE) [63], amyotrophic lateral sclerosis [64] and multiple system atrophy (MSA) [65]. In addition, MSCs have been investigated to be employed in neurological cell dependent therapeutic approaches in adult experimental animals as well as in clinical trials of human brain disorders, as for traumatic brain injury and Huntington's disease [45]. MSCs promote tissue regeneration through production of soluble factors that trigger tissue repair, induce proliferation, migration, and differentiation and help cells, present in the local progenitor, survive. In addition, they decrease inflammatory and immune reactions as well as apoptosis [66, 67]. Transplanted MSCs increase the rate of tissue recovery and host repair through trophic support mediating by direct and indirect mechanisms. Some soluble (cytokines, growth factors) and insoluble (extracellular matrix proteins) factors released by MSCs can enhance the regeneration and survival of the nerve cells through paracrine signaling as well [68]. Other processes that provide specific protection to the brain include remyelination, have been also studied [69]. There are powerful preclinical data that demonstrate that functional recovery through MSCs transplantation can be enhanced following adult hypoxic ischemicor traumatic brain injury [70].

The machineries that govern neural recovery and brain repair by MSCs are being extensively investigated. In cell culture, MSCs can be directed towards neuronal cells and differentiated into mature neurons [71, 72]. MSCs can be stimulated to differentiate under certain culture conditions to cells that possess a Schwann cell morphological features which hold specific markers, as for instance S-100, the nerve growth factor low affinity receptor p75, the glial fibrillary acidic protein (GFAP; a marker of mature astrocytes), and the sulfatide glycolipid O4[73]. Furthermore, they can differentiate *in vitro* into cells with neural nature that have the ability to express neural markers as nestin and neuron-specific nuclear protein (NeuN) [74]. Numerous studies suggested that MSCs secrete soluble factors that can assist in promoting *in vitro* proliferation of neural stem cells or progenitor cells, in addition to being capable of increasing the expression of GFAP [75, 76]. In response to MSCs transplantation that follows brain ischemia, an increase in new oligodendrocyte progenitor cells, mature oligodendrocytes, and formation of myelin in the ischemic hemisphere occurs [77]. One week post experimental cerebral ischemia, human MSCs were transplanted into the injury penumbra of monkeys, which led to reducing the apoptotic cell death, and the volume of the lesion [78].

Additionally, Kim *et al.* [79] reported that administration of human MSCs *via* the intravenous route one day after traumatic brain injury in rats results in the improvement of functional recovery and survival of hostcells by elevating the level of protein kinase B (PKB; also known as pAkt) and reducing the cleavage of caspase-3. MSCs delivered intravenously after one day of mice experimental stroke has also increased axon fiber density, synaptogenesis, and myelination [78]. Among the vital factors of MSCs therapeutic function is their ability to stimulate trophic factors following injury cues, which include Brain-derived neurotrophic factor (BDNF), glial cell derived neurotrophic factor (GDNF), Vascular endothelial growth factor (VEGF), fibroblast growth factor-2, fibronectin, heparin binding-epidermal growth factor-like growth factor (HB-EGF), interleukin-6 (IL-6), Leukemia inhibitory factor (LIF) and platelet-derived growth factor (PDGF) [80]. Not only do MSCs interact directly with the CNS cells, they can also interact with cells of the immune system and can show crucial immuno suppressive properties. MSCs inhibit the proliferation of T- and B-cell, limit the function of natural killer cell, and regulate the microglia and macrophages secretory profile [81]. Consequently, treatment with MSCs is shown to reduce the activity of T- lymphocyte which in turn exerts an immunoregulatory ability [82].

In Ischemia, transplantation of MSCs has been shown to reduce the number of inflammatory cells; activated microglia and macrophages in the brain [83]. By reducing microglial expansion, MSCs transplantation may reduce the inflammatory reaction, which could in turn lead to generating and integrating new neurons. After acute traumatic brain injury in rats, MSCs transplantation could result in regulating the inflammatory reaction by altering the expression of pro- and anti-inflammatory cytokines, in addition to regulating the levels of chemokines in serum [75]. In spinal cord injury of rat model, MSCs transplanted *via* the intravenous route decreased the pro-inflammatory cytokine IL-1ß level and the number of activated microglia, and increased the anti- inflammatory IL-10 level [68]. Moreover, a novel study performed on spinal cord injury showed that, MSCs can stimulate macrophages to change to a M2 anti-inflammatory state from M1 pro-inflammatory phenotype, which promotes a regenerative and protective response [84]. Eventually, regeneration principal in the injured brain relies mainly on angiogenesis and vascular remodeling. MSCs injected *via* the intravenous route in stroke rat model were selectively directed towards ischemic regions of the damaged brain to raise the rate of angiogenesis and provide improvement with the functional recovery [80]. It is evident that these cells produce suitable cytokine milieu required to promote angiogenic response, in addition to producing pro angiogenic factors such as VEGF, and regulate the angiogenesis trophic support such as fibroblast growth factor (bFGF), bone morphogenetic protein2 (BMP-2), angiogenin and IL 6. Intriguingly, MSCs may also possess the ability to differentiate into endothelial cells [85]. The MSCs suggested neuroprotective effects are summarized in Fig. (**1**) [45].

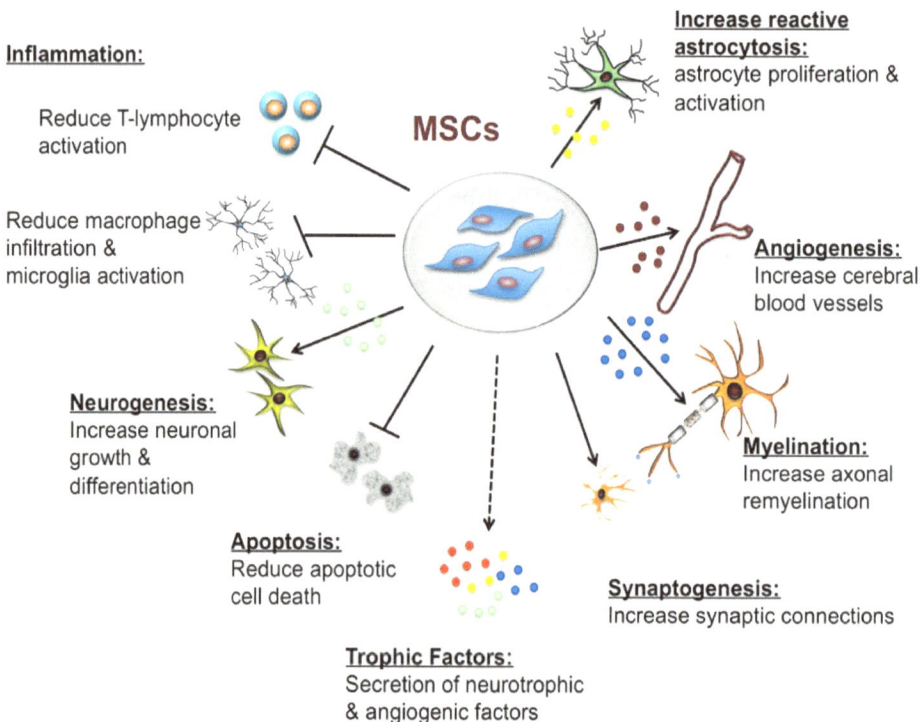

Fig. (1). Potential neuroprotective and neurorestorative effects of MSCs [45].

2.2.4. Adult Stem Cells and Regenerative Medicine

Adult stem cells, or non-ESCs, have been characterized in bone marrow of human and mouse, where they constantly replace the differentiated peripheral blood cells that were lost by attrition. Studies on the haematopoietic system have clarified the stem cell definition as a cell that has the capability to self renew and to develop cells of multitude diversity in the tissue where the stem cell is located. The fact that the haematopoietic stem cells within bone marrow can differentiate to provide all blood elements, has been widely employed in the clinic for the purpose of transplanting bone marrow and stem cells. Reporter assays suggested that some stem cells are capable of differentiating outside their native tissue.

Hematopoietic stem cells are usually driven from the neural tissues [30, 86, 87]. In addition, mesenchymal stem cells present in bone marrow stroma [33, 88], may also develop to glia and neurons [30, 71, 89]. In fact, the great power of the mesenchymal and hematopoietic stem cells of bone marrow deserves to be widely studied and represents a matter of debate [14, 29, 89, 90].

3. STEM CELLS THERAPEUTIC STRATEGY FOR NEURO-DEGENERATIVE DISEASES

Lately, there have been great hopes on the cell replacement therapy, where it has been regarded as an effective therapy that would provide cure to their reversible damaged neuronal systems. Stem cells provided by adult and embryonic tissues and grafted into rats or mice intact brain were most likely followed by their incorporation into the parenchyma of the host and differentiated into functional neural lineages. In the lesioned brain, stem cells could perform migration in the direction of the brain damaged regions, where they could engraft, proliferate and mature into functional neurons. Precursor cells of neural origin can be intravenously administered and can induce functional recovery by migrating into brain damaged areas.

The capacity of CNS is limited for self-repair; therefore various therapeutic strategies, especially stem cell regeneration strategy, are used to repair damaged neurons. The strategy of stem cell regeneration works by replacing cells or providing neuro-protection in the brain disorders.

Stem cells could act as a source of neurons, where they can be implanted to provide replacements for lost cells and circuits in disorders as Parkinson's disease, amyotrophic lateral sclerosis, Huntington's disease, and Alzheimer's disease [12, 30, 34, 91]. Neurodegenerative diseases are identified by the irreparable damage of physiologically functional neurons that takes place on the course of many years, eventually leading to substantial mortality and morbidity [92]. The most common age-dependent neurodegenerative disorders are Alzheimer's and Parkinson's disorders.

3.1. Alzheimer's Disease

Alzheimer's disease (AD) is the most common disorder that results from the dementia and impaired memory that occur at an old age [93]. This terminal and degenerative disease was firstly discovered in 1906 by Alois Alzheimer, the famous German neuropathologist [94]. Alois Alzheimer reported the first case of what was recognized as Alzheimer's disease in a woman called Auguste D., who was fifty years old in age. Alzheimer followed her case until she died in 1906, when he publicly reported the case [95]. AD is usually diagnosed in people whose ages are above 65 years old [96], however, early onset of Alzheimer's can take place but is less prevalent. In 2010, there were 35.6 million afflicted person in the world [87].

Alzheimer's diseaseis projected to globally strike 1 in 85 people by 2050 [97, 98].The main pathological characteristics of Alzheimer's disease are represented in extracellular deposits of beta- amyloid (Aβ) peptide, neuronal loss, neurofibrillary tangles (NFTs) composed of hyperphosphorylated tau, and neurotransmitter dysfunction [93]. These amyloid plaques are equally distributed throughout the hippocampus and the neocortex [99]. A high degree of neuron degeneration takes place on a wide scale in specific susceptible neurons such as the large pyramidal neurons of the inferior temporal cortex, the amygdale, the hippocampus and the entorhinal cortex. Likewise, pyramidal neurons present in the frontal cortex are protected from degeneration [100]. AD has individual differences from one person to another; however, AD patients could share many symptoms. It is wrongly believed that the earliest noticeable symptoms are related to health issues that accompany stress or aging [101].

The primary stages are characterized by being unable to get new memories, as for example suffering with remembering newly observed facts [102]. Body functions are lost, gradually, leading to death, eventually. The mean life expectancy that follows diagnosis ranges from 3 to 9 years [103]. Currently, there are available therapies for AD, such as treatment using acetyl cholinesterase inhibitors that improve the cholinergic function, but provide only temporary and partial symptoms mitigation [104, 105]. Transplantation of cholinergic neurons, that are provided *in vitro* fromnon-stem cell source, could help limit cognitive deterioration caused by the loss of basal forebrain cholinergic neurons. But to afford a long term symptomatic progress, this approach necessitates the presence of intact target cells within the patient's brain, and these are very likely to be harmed. Yet, stem cells have the capacity of transporting factors that can alter the course of the disease, and this is owing to the fact that they possess migratory capacity that follow the transplantation procedure and they can also be genetically modified. The basal forebrain grafts of fibroblasts that release nerve growth factor (NGF), which limit cholinergic neuronal death, induces cell function and improves memory in experimental animals which have been beneficial in AD and this highly supports the approach.

Therefore, cell-based therapy should be capable of differentiating into the three necessary cell types and inducing axonal regeneration and directional synaptogenesis, which would take place by stimulating simultaneous functional and morphological modulation of cells and the environment of the disease. The engrafted cells participate in performing a functional restoration through supplying neurons for signal transmitting and processing, oligodendrocytes for remyelination and neurotrophic factors that act as a defense for the existing host cells [12].

3.2. Parkinson's Disease (PD)

Parkinson's disease (PD) occupies the second place as the most common neurodegenerative disease, accompanied by extrapyramidal motor dysfunction [106]. Parkinson's disease is named after publishing an essay that described in details the shaking palsy, 1817 by James Parkinson [107]. PD is widely spread with elderly, where most cases take place above the age of 50. The primary symptoms of PD include resting tremor, bradykinesia (slowness and decreased amplitude of movement), postural stability and rigidity, and as the disease develops, other symptoms such as dementia, sleep abnormalities, depression and autonomic failure have been identified [108].

PD is characterized by the progressive and selective death neuronal of neuronal heterogeneous populations, which include the neuromelanin-laden dopaminergic neurons of the pars compacta of the substantia nigra, selected aminergic brain-stem nuclei, the cholinergic nucleus basalis of Meynert, hypothalamic neurons, and small cortical neurons, as well as the olfactory bulb, sympathetic ganglia, and parasympathetic neurons in the gut [109]. PD has a specific cell loss pattern and this pattern is reversed in normal aging and is not similar to patterns that characterize the striatonigral degeneration and progressive supranuclear palsy. It leads to regional loss of striatal dopamine, most obviously in the putamen dorsal and intermediate subdivisions [110], a process that is regarded as an interpretation to akinesia and rigidity. Among the crucial pathological features is the presence of degenerating ubiquitin-positive neuronal processes or neuritis, which are present in all affected brain-stem regions, especially the dorsal motor nucleus of the vagus [111].

The proteinacious inclusion bodies known as Lewy bodies and the reactive microgliosis are two pathological entities that characterize the postmortem brains of PD patients [112]. The diagnosis of PD is currently based on the clinical manifestations of the disease [8]. Current data neither hold sufficient information on reliable biomarkers that enable the early detection of PD neurodegeneration, nor do they provide information on detecting and monitoring the impact of drugs on the disease process, but some potential biomarkers exist, such as antibodies that act against neuromelanin, a-synuclein with its pathological forms, DJ-1, and gene expression patterns, protein profiling and metabolomic [113].

Levodopa has been the most extensively used PD therapy over a period of 30 years. The DA precursor L-3,4-dihydroxyphenylalanine (L-DOPA) is converted, by dopa decarboxylase, into dopamine in the dopaminergic neurons [114]. Lack of dopamine in the substantia nigra results in motor symptoms, therefore, administration of L-DOPA temporarily relieves the motor symptoms [112]. A small percentage of L-DOPA (5–10%) passes through the blood-brain barrier. The rest is usually metabolized in a different region to produce dopamine, resulting in a range of side effects that include dyskinesias, joint stiffness and nausea [115]. Carbidopais considered to be an inhibitor of the peripheral dopa decarboxylase, which helps in stopping the L-DOPA metabolism prior to reaching the dopaminergic neurons, which in turn reduces the side effects and increases the bioavailability. It is usually given with levodopa as mixture preparation. The current preparation is called carbidopa/levodopa (co-careldopa, sinemet®) [116].

There is a great interest and a new trend of using human ESCs in order to provide therapeutic approaches that aim at manufacturing a constant supply of dopamine neurons to be implanted into patients [17], where the main aim of this research lies through providing the selective generation of certain cell types to be employed in the regenerative medicine. Scientists have employed a progressive expansion, selection, and differentiation approach to transform mouse ESCs to a different range of mature neurons in tissue culture with 30% of them having dopamine cell features. In addition, authors developed dopamine neurons from mouse ESCs without the formation of embroid body. Clinical trials that involve transplanting human fetal dopaminergic neurons have demonstrated that cell replacement can elicit large, durable improvement in some patients [117]. Therefore, it is worthy of further investigation that cells with properties of dopaminergic neurons have been created *in vitro* from stem cells of many sources, such as ESCs and stem cells derived from bone marrow and fetal brain [118]. However, some PD patients will need implants in many brain regions [119]; to reach optimal recovery. Moreover, it is necessary to develop ways that hamper the progression of the disease [12, 120].

A potential way to avoid the deterioration of native neurons could be by transplanting human stem cells designed to increase the expression of neuroprotective molecules such as glial-cell-line-derived neurotrophic factor (GDNF) [121]. ESCs [122] and NSCs obtained from fetal or adult brain [123] have been applied on a laboratory scale in order to manufacture dopaminergic neurons [12].

Stem cells graft approaches are: pre-differentiation to dopaminergic neurons *in vitro* before transplantation; or differentiation of stem cells after implantation into the striatum or substantia nigra *in vivo*. Fundamentally, the irreversible loss of a single cell phenotype, along with the regular pathology that is specific to PD, proposes treatment strategies that rely on the substitution of this single neuronal cell type. The successful generation of stem cell derived dopamine neurons could provide, in the near future, neurotransplantation for patients suffering from Parkinson's disease.

4. FUTURE PROSPECTIVE

The therapeutic potential of human ESCs and non-ESCs (adult stem cells) has demonstrated for tissues regeneration with high impact value. Basically, endogenous stimulating of stem cells is associated with self-renewal of stem cells that belong to the patient by being administered to particular growth factors. Recently, cell replacement therapy would introduce cure for neurological disorders, for examples AD and PD. ESCs and non- ESCs as well as neurons grafted into the intact brain are considered as promising therapeutic strategy for neuro-associated-diseases. In the lesioned brain, stem cells could perform targeted flow towards the damaged brain regions, where they engrafted, proliferated and developed into functional nerve cells. Functional recovery can occur to brain damaged areas by intravenous administration and migration of neural precursor cells. Hence, stem cell therapy will provide a powerful and effective cure for most of neurodegenrative diseases in near future.

CONFLICT OF INTEREST

The author confirms that author has no conflict of interest to declare for this publication.

ACKNOWLEDGEMENTS

Declared none.

REFERENCES

[1]   Polak J, Mantalaris S, Harding SE. Advances in Tissue Engineering. London: Imperial College Press 2008; pp. 1-903.

[2]   Daar AS, Greenwood HL. A proposed definition of regenerative medicine. J Tissue Eng Regen Med 2007; 1(3): 179-84.
        [http://dx.doi.org/10.1002/term.20] [PMID: 18038409]

[3]   NRC (National Research Council). Stem Cells and the Future of Regenerative Medicine Committee on the Biological and Biomedical Applications of Stem Cell Research, Board on Life Sciences, National Research Council, Board on Neuroscience and Behavioral Health, Institute of Medicine 2002; ISBN: 0-309-50974-2, 112 pages, NATIONAL ACADEMY PRESS, Washington, D.C 2002. Available at: http://www.nap.edu/catalog/10195.html

[4]   Rubinsztein DC. The roles of intracellular protein-degradation pathways in neurodegeneration. Nature 2006; 443(7113): 780-6.
        [http://dx.doi.org/10.1038/nature05291] [PMID: 17051204]

[5]   Jellinger KA. Basic mechanisms of neurodegeneration: a critical update. J Cell Mol Med 2010; 14(3): 457-87.
        [PMID: 20070435]

[6]   Cacciatore I, Baldassarre L, Fornasari E, Mollica A, Pinnen F. Recent advances in the treatment of neurodegenerative diseases based on GSH delivery systems. Oxid Med Cell Longev 2012; 2012: 240146.
        [http://dx.doi.org/10.1155/2012/240146] [PMID: 22701755]

[7]   Farooqui AA, Horrocks LA. Glycerophospholipids in the brain: Phospholipases A2 in neurological disorders. New York: Springer 2007.
        [http://dx.doi.org/10.1007/978-0-387-49931-4]

[8]   Fasano M, Alberio T, Lopiano L. Peripheral biomarkers of Parkinson's disease as early reporters of central neurodegeneration. Biomarkers Med 2008; 2(5): 465-78.
        [http://dx.doi.org/10.2217/17520363.2.5.465] [PMID: 20477424]

[9]   Becker AJ, McCULLOCH EA, Till JE. Cytological demonstration of the clonal nature of spleen colonies derived from transplanted mouse marrow cells. Nature 1963; 197: 452-4.
        [http://dx.doi.org/10.1038/197452a0] [PMID: 13970094]

[10]   Ramakrishna V, Janardhan PB, Sudarsanareddy L. Stem Cells and Regenerative Medicine –A Review. Ann Rev Res Biol 2011; 1: 79-110.

[11]   Avasthi S, Srivastava RN, Singh A, Srivastava M. Stem Cell: Past, Present and Future- A Review Article. IJMU 2008; 3: 22-30.

[12]   Singh P, Williams DJ. Cell therapies: realizing the potential of this new dimension to medical therapeutics. J Tissue Eng Regen Med 2008; 2(6): 307-19.
        [http://dx.doi.org/10.1002/term.108] [PMID: 18618613]

[13]   Lee EH, Hui JH. The potential of stem cells in orthopaedic surgery. J Bone Joint Surg Br 2006; 88(7): 841-51.
        [http://dx.doi.org/10.1302/0301-620X.88B7.17305] [PMID: 16798982]

[14]   Bajada S, Mazakova I, Richardson JB, Ashammakhi N. Updates on stem cells and their applications in regenerative medicine. J Tissue Eng Regen Med 2008; 2(4): 169-83.
        [http://dx.doi.org/10.1002/term.83] [PMID: 18493906]

[15] Evans MJ, Kaufman MH. Establishment in culture of pluripotential cells from mouse embryos. Nature 1981; 292(5819): 154-6.
[http://dx.doi.org/10.1038/292154a0] [PMID: 7242681]

[16] Martin GR. Isolation of a pluripotent cell line from early mouse embryos cultured in medium conditioned by teratocarcinoma stem cells. Proc Natl Acad Sci USA 1981; 78(12): 7634-8.
[http://dx.doi.org/10.1073/pnas.78.12.7634] [PMID: 6950406]

[17] Thomson JA, Itskovitz-Eldor J, Shapiro SS, *et al.* Embryonic stem cell lines derived from human blastocysts. Science 1998; 282(5391): 1145-7.
[http://dx.doi.org/10.1126/science.282.5391.1145] [PMID: 9804556]

[18] Adewumi O, Aflatoonian B, Ahrlund-Richter L, *et al.* Characterization of human embryonic stem cell lines by the International Stem Cell Initiative. Nat Biotechnol 2007; 25(7): 803-16.
[http://dx.doi.org/10.1038/nbt1318] [PMID: 17572666]

[19] Shamblott MJ, Axelman J, Wang S, *et al.* Derivation of pluripotent stem cells from cultured human primordial germ cells. Proc Natl Acad Sci USA 1998; 95(23): 13726-31.
[http://dx.doi.org/10.1073/pnas.95.23.13726] [PMID: 9811868]

[20] Miki T, Lehmann T, Cai H, Stolz DB, Strom SC. Stem cell characteristics of amniotic epithelial cells. Stem Cells 2005; 23(10): 1549-59.
[http://dx.doi.org/10.1634/stem cells.2004-0357] [PMID: 16081662]

[21] De Coppi P, Bartsch G Jr, Siddiqui MM, *et al.* Isolation of amniotic stem cell lines with potential for therapy. Nat Biotechnol 2007; 25(1): 100-6.
[http://dx.doi.org/10.1038/nbt1274] [PMID: 17206138]

[22] Kofidis T, deBruin JL, Yamane T, Balsam LB, *et al.* IGF-1 promotes engraftment, differentiation and functional improvement after transfer of embryonic stem cells for myocardial restoration. Stem Cells 2004; 22: 1239-45.
[http://dx.doi.org/10.1634/stem cells.2004-0127] [PMID: 15579642]

[23] Liu YH, Karra R, Wu SM. Cardiovascular stem cells in regenerative medicine: ready for prime time? Drug Discov Today Ther Strateg 2008; 5(4): 201-7.
[http://dx.doi.org/10.1016/j.ddstr.2008.12.003] [PMID: 20054428]

[24] Frenck RW Jr, Blackburn EH, Shannon KM. The rate of telomere sequence loss in human leukocytes varies with age. Proc Natl Acad Sci USA 1998; 95(10): 5607-10.
[http://dx.doi.org/10.1073/pnas.95.10.5607] [PMID: 9576930]

[25] Rufer N, Brümmendorf TH, Kolvraa S, *et al.* Telomere fluorescence measurements in granulocytes and T lymphocyte subsets point to a high turnover of hematopoietic stem cells and memory T cells in early childhood. J Exp Med 1999; 190(2): 157-67.
[http://dx.doi.org/10.1084/jem.190.2.157] [PMID: 10432279]

[26] Reyes M, Lund T, Lenvik T, Aguiar D, Koodie L, Verfaillie CM. Purification and *ex vivo* expansion of postnatal human marrow mesodermal progenitor cells. Blood 2001; 98(9): 2615-25.
[http://dx.doi.org/10.1182/blood.V98.9.2615] [PMID: 11675329]

[27] Martin MJ, Muotri A, Gage F, Varki A. Human embryonic stem cells express an immunogenic nonhuman sialic acid. Nat Med 2005; 11(2): 228-32.

[http://dx.doi.org/10.1038/nm1181] [PMID: 15685172]

[28]   Landry DW, Zucker HA. Embryonic death and the creation of human embryonic stem cells. J Clin Invest 2004; 114(9): 1184-6.
[http://dx.doi.org/10.1172/JCI23065] [PMID: 15520846]

[29]   Gardner RL. Stem cells and regenerative medicine: principles, prospects and problems. C R Biol 2007; 330(6-7): 465-73.
[http://dx.doi.org/10.1016/j.crvi.2007.01.005] [PMID: 17631439]

[30]   Ma M, Sha C, Zhou Z, Zhou Q, Li Q. Generation of patient-specific pluripotent stem cells and directed differentiation of embryonic stem cells for regenerative medicine. J Nanjing Med Univ 2008; 22: 135-42.
[http://dx.doi.org/10.1016/S1007-4376(08)60052-0]

[31]   Jones R, Lebkowski J, McNiece I. Stem cells. Biol Blood Marrow Transplant 2010; 16(1) (Suppl.): S115-8.
[http://dx.doi.org/10.1016/j.bbmt.2009.10.035] [PMID: 19896546]

[32]   Lagasse E, Connors H, Al-Dhalimy M, *et al.* Purified hematopoietic stem cells can differentiate into hepatocytes *in vivo.* Nat Med 2000; 6(11): 1229-34.
[http://dx.doi.org/10.1038/81326] [PMID: 11062533]

[33]   Abdelkrim H, Juan DB, Jane W, Mohamed A, Bernat S. The immune boundaries for stem cell based therapies: problems and prospective solutions. J Cell Mol Med 13: 1464-75.2009;

[34]   Baizabal JM, Furlan-Magaril M, Santa-Olalla J, Covarrubias L. Neural stem cells in development and regenerative medicine. Arch Med Res 2003; 34(6): 572-88.
[http://dx.doi.org/10.1016/j.arcmed.2003.09.002] [PMID: 14734098]

[35]   Clarke DL, Johansson CB, Wilbertz J, *et al.* Generalized potential of adult neural stem cells. Science 2000; 288(5471): 1660-3.
[http://dx.doi.org/10.1126/science.288.5471.1660] [PMID: 10834848]

[36]   Brignier AC, Gewirtz AM. Embryonic and adult stem cell therapy. J Allergy Clin Immunol 2010; 125(2) (Suppl. 2): S336-44.
[http://dx.doi.org/10.1016/j.jaci.2009.09.032] [PMID: 20061008]

[37]   Si YL, Zhao YL, Hao HJ, Fu XB, Han WD. MSCs: Biological characteristics, clinical applications and their outstanding concerns. Ageing Res Rev 2011; 10(1): 93-103.
[http://dx.doi.org/10.1016/j.arr.2010.08.005] [PMID: 20727988]

[38]   Nauta AJ, Fibbe WE. Immunomodulatory properties of mesenchymal stromal cells. Blood 2007; 110(10): 3499-506.
[http://dx.doi.org/10.1182/blood-2007-02-069716] [PMID: 17664353]

[39]   Oh JY, Kim MK, Shin MS, *et al.* The anti-inflammatory and anti-angiogenic role of mesenchymal stem cells in corneal wound healing following chemical injury. Stem Cells 2008; 26(4): 1047-55.
[http://dx.doi.org/10.1634/stem cells.2007-0737] [PMID: 18192235]

[40]   Mahmood A, Lu D, Chopp M. Intravenous administration of marrow stromal cells (MSCs) increases the expression of growth factors in rat brain after traumatic brain injury. J Neurotrauma 2004; 21(1): 33-9.

[http://dx.doi.org/10.1089/089771504772695922] [PMID: 14987463]

[41]   Lu D, Mahmood A, Wang L, Li Y, Lu M, Chopp M. Adult bone marrow stromal cells administered intravenously to rats after traumatic brain injury migrate into brain and improve neurological outcome. Neuroreport 2001; 12(3): 559-63.
[http://dx.doi.org/10.1097/00001756-200103050-00025] [PMID: 11234763]

[42]   Brehm M, Zeus T, Kostering M, Kogler G, Wernet P, Strauer BE. Intracoronary transplantation of autologous bone-marrow cells for therapeutic angiogenesis in patients with myocardial infarction. Eur J Clin Invest 2002; 32(2): 77.

[43]   da Silva Meirelles L, Caplan AI, Nardi NB. In search of the *in vivo* identity of mesenchymal stem cells. Stem Cells 2008; 26(9): 2287-99.
[http://dx.doi.org/10.1634/stem cells.2007-1122] [PMID: 18566331]

[44]   Tögel F, Weiss K, Yang Y, Hu Z, Zhang P, Westenfelder C. Vasculotropic, paracrine actions of infused mesenchymal stem cells are important to the recovery from acute kidney injury. Am J Physiol Renal Physiol 2007; 292(5): F1626-35.
[http://dx.doi.org/10.1152/ajprenal.00339.2006] [PMID: 17213465]

[45]   Castillo-Melendez M, Yawno T, Jenkin G, Miller SL. Stem cell therapy to protect and repair the developing brain: a review of mechanisms of action of cord blood and amnion epithelial derived cells. Front Neurosci 2013; 7: 194.
[http://dx.doi.org/10.3389/fnins.2013.00194] [PMID: 24167471]

[46]   Dharmasaroja P. Bone marrow-derived mesenchymal stem cells for the treatment of ischemic stroke. J Clin Neurosci 2009; 16(1): 12-20.
[http://dx.doi.org/10.1016/j.jocn.2008.05.006] [PMID: 19017556]

[47]   Ye M, Wang XJ, Zhang YH, *et al.* Therapeutic effects of differentiated bone marrow stromal cell transplantation on rat models of Parkinson's disease. Parkinsonism Relat Disord 2007; 13(1): 44-9.
[http://dx.doi.org/10.1016/j.parkreldis.2006.07.013] [PMID: 17005432]

[48]   de Girolamo L, Lucarelli E, Alessandri G, *et al.* Mesenchymal stem/stromal cells: a new "cells as drugs" paradigm. Efficacy and critical aspects in cell therapy. Curr Pharm Des 2013; 19(13): 2459-73.
[http://dx.doi.org/10.2174/1381612811319130015] [PMID: 23278600]

[49]   Li L, Xia Y. Study of adipose tissue-derived mesenchymal stem cells transplantation for rats with dilated cardiomyopathy. Ann Thorac Cardiovasc Surg 2014; 20(5): 398-406. [Epub ahead of print].
[http://dx.doi.org/10.5761/atcs.oa.13-00104] [PMID: 24492176]

[50]   Mizuno H. Adipose-derived stem cells for tissue repair and regeneration: ten years of research and a literature review. J Nippon Med Sch 2009; 76(2): 56-66.
[http://dx.doi.org/10.1272/jnms.76.56] [PMID: 19443990]

[51]   Levi B, James AW, Nelson ER, *et al.* Human adipose derived stromal cells heal critical size mouse calvarial defects. PLoS One 2010; 5(6): e11177.
[http://dx.doi.org/10.1371/journal.pone.0011177] [PMID: 20567510]

[52]   Markov V, Kusumi K, Tadesse MG, *et al.* Identification of cord blood-derived mesenchymal stem/stromal cell populations with distinct growth kinetics, differentiation potentials, and gene expression profiles. Stem Cells Dev 2007; 16(1): 53-73.

[http://dx.doi.org/10.1089/scd.2006.0660] [PMID: 17348805]

[53]   Alviano F, Fossati V, Marchionni C, *et al.* Term Amniotic membrane is a high throughput source for multipotent Mesenchymal Stem Cells with the ability to differentiate into endothelial cells *in vitro.* BMC Dev Biol 2007; 7: 11.
[http://dx.doi.org/10.1186/1471-213X-7-11] [PMID: 17313666]

[54]   Chang YJ, Hwang SM, Tseng CP, *et al.* Isolation of mesenchymal stem cells with neurogenic potential from the mesoderm of the amniotic membrane. Cells Tissues Organs (Print) 2010; 192(2): 93-105.
[http://dx.doi.org/10.1159/000295774] [PMID: 20215735]

[55]   Chavakis E, Urbich C, Dimmeler S. Homing and engraftment of progenitor cells: a prerequisite for cell therapy. J Mol Cell Cardiol 2008; 45(4): 514-22.
[http://dx.doi.org/10.1016/j.yjmcc.2008.01.004] [PMID: 18304573]

[56]   Chapel A, Bertho JM, Bensidhoum M, *et al.* Mesenchymal stem cells home to injured tissues when co-infused with hematopoietic cells to treat a radiation-induced multi-organ failure syndrome. J Gene Med 2003; 5(12): 1028-38.
[http://dx.doi.org/10.1002/jgm.452] [PMID: 14661178]

[57]   Brooke G, Tong H, Levesque JP, Atkinson K. Molecular trafficking mechanisms of multipotent mesenchymal stem cells derived from human bone marrow and placenta. Stem Cells Dev 2008; 17(5): 929-40.
[http://dx.doi.org/10.1089/scd.2007.0156] [PMID: 18564033]

[58]   Le Blanc K, Ringdén O. Immunomodulation by mesenchymal stem cells and clinical experience. J Intern Med 2007; 262(5): 509-25.
[http://dx.doi.org/10.1111/j.1365-2796.2007.01844.x] [PMID: 17949362]

[59]   Cselenyák A, Pankotai E, Horváth EM, Kiss L, Lacza Z. Mesenchymal stem cells rescue cardiomyoblasts from cell death in an *in vitro* ischemia model *via* direct cell-to-cell connections. BMC Cell Biol 2010; 11: 29.
[http://dx.doi.org/10.1186/1471-2121-11-29] [PMID: 20406471]

[60]   Ponte AL, Marais E, Gallay N, *et al.* The *in vitro* migration capacity of human bone marrow mesenchymal stem cells: comparison of chemokine and growth factor chemotactic activities. Stem Cells 2007; 25(7): 1737-45.
[http://dx.doi.org/10.1634/stem cells.2007-0054] [PMID: 17395768]

[61]   Himes BT, Neuhuber B, Coleman C, *et al.* Recovery of function following grafting of human bone marrow-derived stromal cells into the injured spinal cord. Neurorehabil Neural Repair 2006; 20(2): 278-96.
[http://dx.doi.org/10.1177/1545968306286976] [PMID: 16679505]

[62]   Park HJ, Lee PH, Bang OY, Lee G, Ahn YH. Mesenchymal stem cells therapy exerts neuroprotection in a progressive animal model of Parkinson's disease. J Neurochem 2008; 107(1): 141-51.
[http://dx.doi.org/10.1111/j.1471-4159.2008.05589.x] [PMID: 18665911]

[63]   Zhang J, Li Y, Lu M, *et al.* Bone marrow stromal cells reduce axonal loss in experimental autoimmune encephalomyelitis mice. J Neurosci Res 2006; 84(3): 587-95.
[http://dx.doi.org/10.1002/jnr.20962] [PMID: 16773650]

[64] Choi MR, Kim HY, Park JY, *et al.* Selection of optimal passage of bone marrow-derived mesenchymal stem cells for stem cell therapy in patients with amyotrophic lateral sclerosis. Neurosci Lett 2010; 472(2): 94-8.
[http://dx.doi.org/10.1016/j.neulet.2010.01.054] [PMID: 20117176]

[65] Lee PH, Park HJ. Bone marrow-derived mesenchymal stem cell therapy as a candidate disease-modifying strategy in Parkinson's disease and multiple system atrophy. J Clin Neurol 2009; 5(1): 1-10.
[http://dx.doi.org/10.3988/jcn.2009.5.1.1] [PMID: 19513327]

[66] Caplan AI, Dennis JE. Mesenchymal stem cells as trophic mediators. J Cell Biochem 2006; 98(5): 1076-84.
[http://dx.doi.org/10.1002/jcb.20886] [PMID: 16619257]

[67] Yi ES, Lee SH, Son MH, *et al.* Hematopoietic stem cell transplantation in children with acute leukemia: similar outcomes in recipients of umbilical cord blood *versus* marrow or peripheral blood stem cells from related or unrelated donors. Korean J Pediatr 2012; 55(3): 93-9.
[http://dx.doi.org/10.3345/kjp.2012.55.3.93] [PMID: 22474464]

[68] Seo JH, Cho SR. Neurorestoration induced by mesenchymal stem cells: potential therapeutic mechanisms for clinical trials. Yonsei Med J 2012; 53(6): 1059-67.
[http://dx.doi.org/10.3349/ymj.2012.53.6.1059] [PMID: 23074102]

[69] Akiyama K, Chen C, Wang D, *et al.* Mesenchymal-stem-cell-induced immunoregulation involves FAS-ligand-/FAS-mediated T cell apoptosis. Cell Stem Cell 2012; 10(5): 544-55.
[http://dx.doi.org/10.1016/j.stem.2012.03.007] [PMID: 22542159]

[70] Parr AM, Tator CH, Keating A. Bone marrow-derived mesenchymal stromal cells for the repair of central nervous system injury. Bone Marrow Transplant 2007; 40(7): 609-19.
[http://dx.doi.org/10.1038/sj.bmt.1705757] [PMID: 17603514]

[71] Woodbury D, Schwarz EJ, Prockop DJ, Black IB. Adult rat and human bone marrow stromal cells differentiate into neurons. J Neurosci Res 2000; 61(4): 364-70.
[http://dx.doi.org/10.1002/1097-4547(20000815)61:4<364::AID-JNR2>3.0.CO;2-C] [PMID: 10931522]

[72] Deng J, Petersen BE, Steindler DA, Jorgensen ML, Laywell ED. Mesenchymal stem cells spontaneously express neural proteins in culture and are neurogenic after transplantation. Stem Cells 2006; 24(4): 1054-64.
[http://dx.doi.org/10.1634/stem cells.2005-0370] [PMID: 16322639]

[73] Zhang J, Li Y, Chen J, *et al.* Expression of insulin-like growth factor 1 and receptor in ischemic rats treated with human marrow stromal cells. Brain Res 2004; 1030(1): 19-27.
[http://dx.doi.org/10.1016/j.brainres.2004.09.061] [PMID: 15567334]

[74] Sanchez-Ramos J, Song S, Cardozo-Pelaez F, *et al.* Adult bone marrow stromal cells differentiate into neural cells *in vitro.* Exp Neurol 2000; 164(2): 247-56.
[http://dx.doi.org/10.1006/exnr.2000.7389] [PMID: 10915564]

[75] Galindo LT, Filippo TR, Semedo P, *et al.* Mesenchymal stem cell therapy modulates the inflammatory response in experimental traumatic brain injury. Neurol Res Int 2011.

[76] Munoz JL, Greco SJ, Patel SA, *et al.* Feline bone marrow-derived mesenchymal stromal cells (MSCs)

show similar phenotype and functions with regards to neuronal differentiation as human MSCs. Differentiation 2012; 84(2): 214-22.
[http://dx.doi.org/10.1016/j.diff.2012.07.002] [PMID: 22824626]

[77]   van Velthoven CT, Kavelaars A, van Bel F, Heijnen CJ. Mesenchymal stem cell treatment after neonatal hypoxic-ischemic brain injury improves behavioral outcome and induces neuronal and oligodendrocyte regeneration. Brain Behav Immun 2010; 24(3): 387-93.
[http://dx.doi.org/10.1016/j.bbi.2009.10.017] [PMID: 19883750]

[78]   Xin H, Li Y, Shen LH, *et al.* Increasing tPA activity in astrocytes induced by multipotent mesenchymal stromal cells facilitate neurite outgrowth after stroke in the mouse. PLoS One 2010; 5(2): e9027.
[http://dx.doi.org/10.1371/journal.pone.0009027] [PMID: 20140248]

[79]   Kim HJ, Lee JH, Kim SH. Therapeutic effects of human mesenchymal stem cells on traumatic brain injury in rats: secretion of neurotrophic factors and inhibition of apoptosis. J Neurotrauma 2010; 27(1): 131-8.
[http://dx.doi.org/10.1089/neu.2008.0818] [PMID: 19508155]

[80]   Li SF, Lu XF, Sun MH. Biological characteristics of mesenchymal stem cells *in vitro* derived from bone marrow of banna minipig inbred line. Zhongguo Xiu Fu Chong Jian Wai Ke Za Zhi 2002; 16(5): 354-8.
[PMID: 12569813]

[81]   Le Blanc K. Immunomodulatory effects of fetal and adult mesenchymal stem cells. Cytotherapy 2003; 5(6): 485-9.
[http://dx.doi.org/10.1080/14653240310003611] [PMID: 14660044]

[82]   Fibbe WE, Nauta AJ, Roelofs H. Modulation of immune responses by mesenchymal stem cells. Ann N Y Acad Sci 2007; 1106: 272-8.
[http://dx.doi.org/10.1196/annals.1392.025] [PMID: 17442776]

[83]   Kim D, Chun BG, Kim YK, *et al. In vivo* tracking of human mesenchymal stem cells in experimental stroke. Cell Transplant 2008; 16(10): 1007-12.
[http://dx.doi.org/10.3727/000000007783472381] [PMID: 18351016]

[84]   Busch SA, Hamilton JA, Horn KP, *et al.* Multipotent adult progenitor cells prevent macrophage-mediated axonal dieback and promote regrowth after spinal cord injury. J Neurosci 2011; 31(3): 944-53.
[http://dx.doi.org/10.1523/JNEUROSCI.3566-10.2011] [PMID: 21248119]

[85]   Potapova IA, Gaudette GR, Brink PR, *et al.* Mesenchymal stem cells support migration, extracellular matrix invasion, proliferation, and survival of endothelial cells *in vitro.* Stem Cells 2007; 25(7): 1761-8.
[http://dx.doi.org/10.1634/stem cells.2007-0022] [PMID: 17395769]

[86]   Jackson KA, Mi T, Goodell MA. Hematopoietic potential of stem cells isolated from murine skeletal muscle. Proc Natl Acad Sci USA 1999; 96(25): 14482-6.
[http://dx.doi.org/10.1073/pnas.96.25.14482] [PMID: 10588731]

[87]   Galli R, Borello U, Gritti A, *et al.* Skeletal myogenic potential of human and mouse neural stem cells. Nat Neurosci 2000; 3(10): 986-91.

[http://dx.doi.org/10.1038/79924] [PMID: 11017170]

[88]   Liechty KW, MacKenzie TC, Shaaban AF, *et al.* Human mesenchymal stem cells engraft and demonstrate site-specific differentiation after in utero transplantation in sheep. Nat Med 2000; 6(11): 1282-6.
[http://dx.doi.org/10.1038/81395] [PMID: 11062543]

[89]   Mezey E, Chandross KJ, Harta G, Maki RA, McKercher SR. Turning blood into brain: cells bearing neuronal antigens generated *in vivo* from bone marrow. Science 2000; 290(5497): 1779-82.
[http://dx.doi.org/10.1126/science.290.5497.1779] [PMID: 11099419]

[90]   Weissman IL. Stem cells: units of development, units of regeneration, and units in evolution. Cell 2000; 100(1): 157-68.
[http://dx.doi.org/10.1016/S0092-8674(00)81692-X] [PMID: 10647940]

[91]   Sheyn D, Mizrahi O, Benjamin S, Gazit Z, Pelled G, Gazit D. Genetically modified cells in regenerative medicine and tissue engineering. Adv Drug Deliv Rev 2010; 62(7-8): 683-98.
[http://dx.doi.org/10.1016/j.addr.2010.01.002] [PMID: 20114067]

[92]   Steiner AF, Karl N, Pilapil C, Noth U, Evans CH, Murray MM. Multilineage mesenchymal differentiation potential of cells migrating out of the anterior cruciate ligament. Transactions of the 52nd Annual Meeting of the Orthopaedic Research Society Chicago IL. 1133.

[93]   Tayeb HO, Yang HD, Price BH, Tarazi FI. Pharmacotherapies for Alzheimer's disease: beyond cholinesterase inhibitors. Pharmacol Ther 2012; 134(1): 8-25.
[http://dx.doi.org/10.1016/j.pharmthera.2011.12.002] [PMID: 22198801]

[94]   Berchtold NC, Cotman CW. Evolution in the conceptualization of dementia and Alzheimer's disease: Greco-Roman period to the 1960s. Neurobiol Aging 1998; 19(3): 173-89.
[http://dx.doi.org/10.1016/S0197-4580(98)00052-9] [PMID: 9661992]

[95]   Ulrike M, Konrad M. Alzheimer: the life of a physician and the career of a disease. Columbia University Press New York 2003; p. 270.

[96]   Mayeux R. Epidemiology of neurodegeneration. Annu Rev Neurosci 2003; 26: 81-104.
[http://dx.doi.org/10.1146/annurev.neuro.26.043002.094919] [PMID: 12574495]

[97]   Prince M, Bryce R, Albanese E, Wimo A, Ribeiro W, Ferri CP. The global prevalence of dementia: A systematic review and metaanalysis. Alzheimers Dement. J Alzheimers Assoc 2013; 9: 63-75.

[98]   Rocca WA, Petersen RC, Knopman DS, *et al.* Trends in the incidence and prevalence of Alzheimer's disease, dementia, and cognitive impairment in the United States. Alzheimers Dement. J Alzheimers Assoc 2011; 7: 80-93.

[99]   Arnold SE, Hyman BT, Flory J, Damasio AR, Van Hoesen GW. The topographical and neuroanatomical distribution of neurofibrillary tangles and neuritic plaques in the cerebral cortex of patients with Alzheimer's disease. Cereb Cortex 1991; 1(1): 103-16.
[http://dx.doi.org/10.1093/cercor/1.1.103] [PMID: 1822725]

[100]   Gómez-Isla T, Hollister R, West H, *et al.* Neuronal loss correlates with but exceeds neurofibrillary tangles in Alzheimer's disease. Ann Neurol 1997; 41(1): 17-24.
[http://dx.doi.org/10.1002/ana.410410106] [PMID: 9005861]

[101] Waldemar G, Dubois B, Emre M, *et al.* Recommendations for the diagnosis and management of Alzheimer's disease and other disorders associated with dementia: EFNS guideline. Eur J Neurol 2007; 14(1): e1-e26.
[http://dx.doi.org/10.1111/j.1468-1331.2006.01605.x] [PMID: 17222085]

[102] Tabert MH, Liu X, Doty RL, *et al.* A 10-item smell identification scale related to risk for Alzheimer's disease. Ann Neurol 2005; 58(1): 155-60.
[http://dx.doi.org/10.1002/ana.20533] [PMID: 15984022]

[103] Querfurth HW, LaFerla FM. Alzheimer's disease. N Engl J Med 2010; 362(4): 329-44.
[http://dx.doi.org/10.1056/NEJMra0909142] [PMID: 20107219]

[104] Birks J. Cholinesterase inhibitors for Alzheimer's disease. Cochrane Database Syst Rev 2006; 1(1): CD005593.
[PMID: 16437532]

[105] Lanctôt KL, Rajaram RD, Herrmann N. Review: Therapy for Alzheimer's disease: How effective are current treatments? Ther Adv Neurol Disorder 2009; 2(3): 163-80.
[http://dx.doi.org/10.1177/1756285609102724] [PMID: 21179526]

[106] Olanow CW, Perl DP, DeMartino GN, McNaught KS. Lewy-body formation is an aggresome-related process: a hypothesis. Lancet Neurol 2004; 3(8): 496-503.
[http://dx.doi.org/10.1016/S1474-4422(04)00827-0] [PMID: 15261611]

[107] Parkinson J. An essay on the shaking palsy. 1817. J Neuropsychiatry Clin Neurosci 2002; 14(2): 223-36.
[http://dx.doi.org/10.1176/jnp.14.2.223] [PMID: 11983801]

[108] Chaudhuri KR, Schapira AH. Non-motor symptoms of Parkinson's disease: dopaminergic pathophysiology and treatment. Lancet Neurol 2009; 8(5): 464-74.
[http://dx.doi.org/10.1016/S1474-4422(09)70068-7] [PMID: 19375664]

[109] Fearnley JM, Lees AJ. Ageing and Parkinson's disease: substantia nigra regional selectivity. Brain 1991; 114(Pt 5): 2283-301.
[http://dx.doi.org/10.1093/brain/114.5.2283] [PMID: 1933245]

[110] Kish SJ, Shannak K, Hornykiewicz O. Uneven pattern of dopamine loss in the striatum of patients with idiopathic Parkinson's disease. Pathophysiologic and clinical implications. N Engl J Med 1988; 318(14): 876-80.
[http://dx.doi.org/10.1056/NEJM198804073181402] [PMID: 3352672]

[111] Gai WP, Blessing WW, Blumbergs PC. Ubiquitin-positive degenerating neurites in the brainstem in Parkinson's disease. Brain 1995; 118(Pt 6): 1447-59.
[http://dx.doi.org/10.1093/brain/118.6.1447] [PMID: 8595476]

[112] Kim YS, Kim YK, Hwang O, Kim DJ. Pathology of neurodegenerative diseases, Brain damage - Bridging between basic research and clinics . , InTech 2012; pp. 99-138.

[113] Gerlach M, Maetzler W, Broich K, *et al.* Biomarker candidates of neurodegeneration in Parkinson's disease for the evaluation of disease-modifying therapeutics. J Neural Transm (Vienna) 2012; 119(1): 39-52.
[http://dx.doi.org/10.1007/s00702-011-0682-x] [PMID: 21755462]

[114] Nagal A, Singla RK. Parkinson's disease: diagnosis, therapeutics and management. Webmed Central Pharmaceutic Sci 2012; 3

[115] Kostic V, Przedborski S, Flaster E, Sternic N. Early development of levodopa-induced dyskinesias and response fluctuations in young-onset Parkinson's disease. Neurology 1991; 41(2 (Pt 1)): 202-5.
[http://dx.doi.org/10.1212/WNL.41.2_Part_1.202] [PMID: 1992362]

[116] Abdel-Salam OM. Drugs used to treat Parkinson's disease, present status and future directions. CNS Neurol Disord Drug Targets 2008; 7(4): 321-42.
[http://dx.doi.org/10.2174/187152708786441867] [PMID: 18991661]

[117] Lindvall O, Kokaia Z, Martinez-Serrano A. Stem cell therapy for human neurodegenerative disorders-how to make it work. Nat Med 2004; 10 (Suppl.): S42-50.
[http://dx.doi.org/10.1038/nm1064] [PMID: 15272269]

[118] Takagi Y, Takahashi J, Saiki H, *et al.* Dopaminergic neurons generated from monkey embryonic stem cells function in a Parkinson primate model. J Clin Invest 2005; 115(1): 102-9.
[http://dx.doi.org/10.1172/JCI21137] [PMID: 15630449]

[119] Piccini P, Pavese N, Hagell P, *et al.* Factors affecting the clinical outcome after neural transplantation in Parkinson's disease. Brain 2005; 128(Pt 12): 2977-86.
[http://dx.doi.org/10.1093/brain/awh649] [PMID: 16246865]

[120] Anderson WS, Lenz FA. Surgery insight: Deep brain stimulation for movement disorders. Nat Clin Pract Neurol 2006; 2(6): 310-20.
[http://dx.doi.org/10.1038/ncpneuro0193] [PMID: 16932575]

[121] Behrstock S, Ebert A, McHugh J, *et al.* Human neural progenitors deliver glial cell line-derived neurotrophic factor to parkinsonian rodents and aged primates. Gene Ther 2006; 13(5): 379-88.
[http://dx.doi.org/10.1038/sj.gt.3302679] [PMID: 16355116]

[122] Ho HY, Li M. Potential application of embryonic stem cells in Parkinson's disease: drug screening and cell therapy. Regen Med 2006; 1(2): 175-82.
[http://dx.doi.org/10.2217/17460751.1.2.175] [PMID: 17465801]

[123] Sanberg PR. Neural stem cells for Parkinson's disease: to protect and repair. Proc Natl Acad Sci USA 2007; 104(29): 11869-70.
[http://dx.doi.org/10.1073/pnas.0704704104] [PMID: 17620601]

Developmental and Stem Cell Biology in Health and Disease, 2016, 193-209   **193**

CHAPTER 10

Stem Cells and Bone Disorders

Hanaa H. Ahmed[*, 1], **Sara M. Abdo**[2] and **Ahmed A. Abd-Rabou**[1]

[1] *Hormones Department, Medical Research Division, National Research Center, Cairo, Egypt*

[2] *Chemistry Department, Faculty of Science, Helwan University, Egypt*

Abstract: The composition of bone is a connective tissue of particular cells, fibers, and ground substance. One of the well known bone diseases, which featured by low bone mass and bone fragility, is osteoporosis. Nowadays, there is a fast ongoing revolution in stem cells either derived from blood or tissues postnatally. The present chapter has provided a principal understanding about the importance of mesenchymal stem cells (MSCs) in the management of osteoporosis. This might be attributed to their direct ability to generate osteoprogenitors and osteoblasts *via* their influence on osteoclastogenesis. In addition, the usefulness of the MSCs when combined with calcium phosphate composite as an osteoinductive material in osteoporosis stem cell therapy has been described in the current chapter. On the other hand, self-renewal dental pulp stem cells are now known as being important to the regeneration of dentine. Development of novel tissue engineering strategies is mainly based on excellent awareness of the stem cells nature to determine their potentialities. This novel way of bone disease therapy may provide an innovative generation of new strategies to tackle bone diseases.

Keywords: Bone, Osteoporosis, Osteoinductive materials, Stem cells.

1. INTRODUCTION

Osteoporosis, a bone disease, is featured by a bone fracture rate increment, which causes considerable morbidity and mortality [1]. Bone fracture cases related to osteoporosis is expected to be doubled by 2050 [2]. These fracture incidences are increasing particularly among the elderly population in the developed nations [1].

---

[*] **Correspondence author Hanaa H. Ahmed:** Medical Research Division, National Research Center, 12622, Cairo, Egypt; Tel: +2 01223661935; Fax: 02-33370931; E-mail: hanaaomr@yahoo.com

Ahmed El-Hashash (Ed)

In many cases, osteoporosis is only diagnosed after the first clinical bone fracture, due to bone loss occurs artfully as well as it is asymptomatic. In light of these evidences, early diagnosis of osteoporosis is very important to prevent the first fracture and its therapy is vital for further fracture protection [3]. White and old females with decreased calcium intake, decreased estrogen secretion, thin body, alcohol abuse, and cigarette smoking are highly susceptible for osteoporosis. Moreover, some genetic factors were investigated to be considered as risk for osteoporosis incidences [4].

Currently, some interventional strategies for osteoporosis were reported such as receptor activator of NF-κB ligand inhibitors, cathepsin K inhibitors, selective estrogen receptor modulators (SERM), sclerostin (SOST), and parathyroid hormone inhibitors [5]. It was reported that genetic transfection of bone morphogenic protein-2 (BMP-2) using bone tissue engineering is a powerful technique for bone regeneration [6], where it was used as a therapeutic strategy in tackling bone fractures; however, the high cost is remaining a main obstacle in applying such this approach. Preclinical studies showed moderate burden effects such as ectopic bone resorption and formation, hematoma, and minor immunologic reactions [7].

Osteoblasts are very important cells in bone reconstruction. Development of mesenchymal stem cells (MSCs) is associated with mesodermal sclerotome condensation that subsequently differentiates to other cell types for bone remodeling. Some stimulatory factors were needed for MSCs derived from osteoblastogenesis such as BMPs, which act as the starting inducer of osteoblastogenesis, as well as members of the transforming growth factor β (TGFβ) superfamily. This interesting story is involving in fracture tackling. On the other hand, MSCs derived from bone marrow fat is another vital story in accelerating bone fracture tackling [8]. This approach was used for osteogenesis imperfect therapy [9].

The natural regenerative effect of dentine/pulp complex leads to tertiary dentine formation. Odontoblasts may survive mild injury and secrete the dentine matrix. However, trauma of advanced intensity may lead to the pre-existing cell death. In response to particular stimuli, new odontoblasts are recruited to the injury and differentiated. This reparative mechanism plays an important role in preserving pulp vitality [10, 11]. This chapter aimed at considering the stem cells and related technologies in the future of the clinical applications.

2. BONE STRUCTURE

Bone is a specialized connective tissue, unlike others, where its extracellular components are mineralized providing it with strength and rigidity. This makes bone identically suited to fulfilling its vital role of body mechanical supports and necessary locomotion process [12]. It also provides a support for soft tissues, represents points of attachment for skeletal muscles, protects internal organs, including bone marrow, as well as it plays central role in mineral homeostasis, principally of calcium and phosphorus [13, 14].

Osteon is the building unit of bone; consisting of concentric lamellae surrounding a Haversian canal. Osteons are basically separated by interstitial lamellae, formed from osteon remnants [14]. The smooth, white, and solid cortical bone tissue is the hard outer layer of bones and it is approximately 80% of skeletal tissue mass [15].

Trabecular bone, the hole-filled spongy bone tissue, has high pores covering 50-90 % of its volume. The porous network are spreading vertically and horizontally, giving the trabecular bone in a sponge appearance [12, 15].

The periosteum covers external bone surfaces and is classified into two particular layers; an outer fibrous and inner cellular (cambium) layers. The ambium layer is a very important layer located in a direct contact with the bone surface and it contains MSCs that have the potentiality for differentiation into osteoblasts, chondrocytes, and progenitor cells. Bone structure is illustrated in Fig. (**1**) [12].

2.1. Bone Matrix

Bone matrix is consisting of organic and inorganic partitions. Type I collagen (approximately 95%), which provides bone with its flexibility, is the main component of the organic part. The remainder is made up of proteoglycans and noncollagenous proteins including osteocalcin, osteonectin, and osteopontin. On the other hand, the inorganic partition of the bone matrix, which acts as an ion reservoir, represents approximately 65-70%. Bone stiffness is featured by the crystalline structures of the ions, for example calcium hydroxyapatite [Ca_{10} $(PO_4)_6$ $(OH)_2$], fluoride, sodium, potassium, and magnesium. Collagen and mineral distribution provides bone with its stiffness and flexibility [12,14].

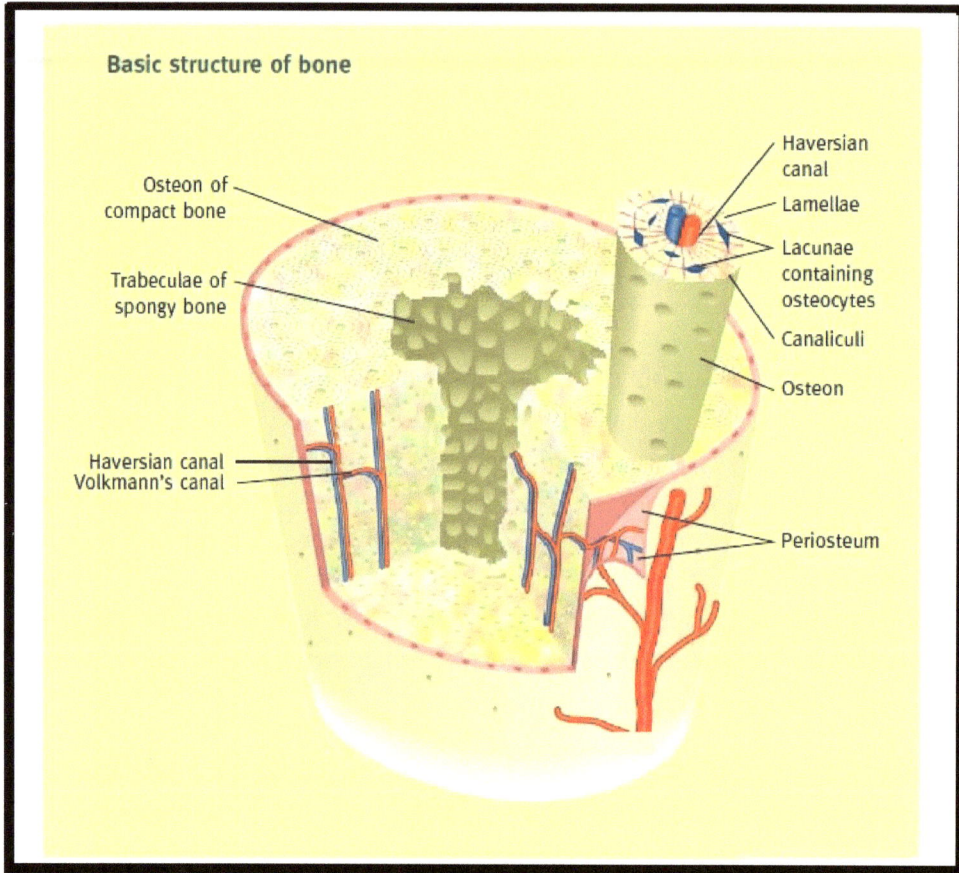

Fig. (1). Bone structure [14].

2.2. Cellular Elements

The major cell types of bone are osteoblasts (OBs), osteoclasts (OCs), and osteocytes [16]. The haematopoietic stem cells (HSCs) and mesenchymal stem cells give rise to the osteoclasts that modulate bone resorption and osteoblasts that mediate bone formation [12].

2.3. Bone Remodeling

Bone remodeling is a skeletal dynamic process based on Basic Multicellular Unit (BMU), which consists of osteoclasts, osteoblasts, and osteocytes [17]. In the BMU, the amount of bone destroyed by osteoclasts is exactly equal to the amount of bone formed by osteoblasts [18]. Moreover, bone remodeling is a life-dominant process, where cortical and trabecular bone is rebuilt, and it plays a crucial role in bone mass balance and mineral homeostasis [19]. Intriguingly, non-BMU cells, including T-cells, B-cells, and nerve cells also play an important role in bone remodeling [20 - 22].

2.3.1. Mechanism of Bone Remodeling

Initiation phase is the first process in the remodeling cycle; where osteoclast precursors are recruiting and differentiated into mature osteoclasts to maintain bone resorption. A *reversal period* is associated with bone resorption inhibition and the osteoblasts are recruited and started to be differentiated. *Termination phase* is the final step in the cycle; where bone formation is proceeding by the osteoblasts [17, 23]. Finally, there is a connected network exists among each of those cells and the osteocytes [17].

The *termination phase* is based on a mechanism that requires a key mediator called receptor activator of nuclear factor kappa-B ligand (RANKL), which is essentially required for osteoclastogenesis [24, 25]. Intriguingly, ligand-receptor interaction of RANKL-RANK triggers the *termination phase* by promoting osteoclast precursor fusion and colony stimulating factor (CSF-1) [26 - 29].

During the remodeling process, the majority of bone turnover markers, that quantify bone resorption, are collagen breakdown products, telopeptides, and pyridinoline rings [30]. Deoxypyridinoline (DPD) is released into the circulation after resorption of bone matrix and thereafter excreted in urine [31]. BMU at the initiation, transition and termination phases of bone remodeling cycle is illustrated in Fig. (**2**).

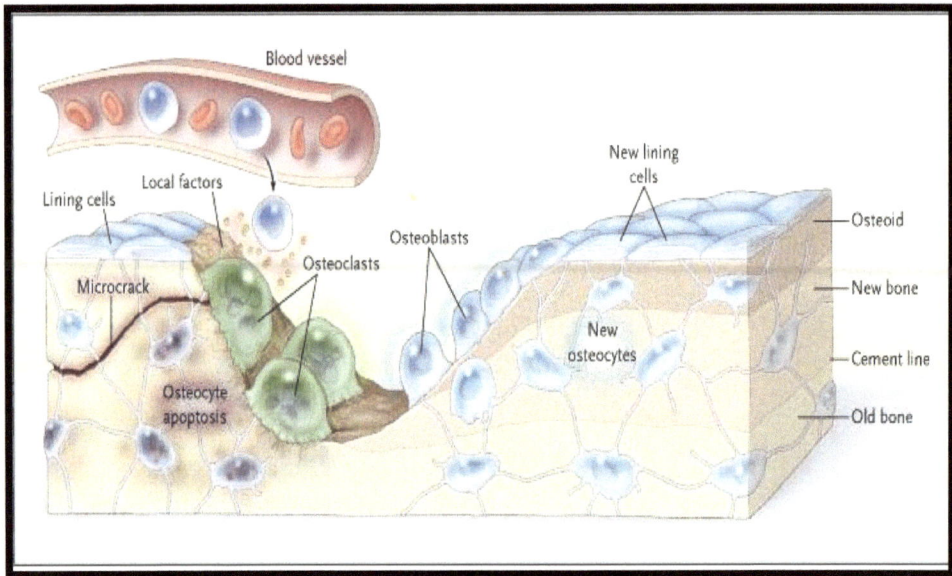

Fig. (2). BMU at the initiation, transition and termination phases of bone remodeling cycle [23].

3. OSTEOPOROSIS

Bone fragility and fracture are the major diagnostic markers for osteoporosis. A more scientific definition of osteoporosis is associated with bone mineral density (BMD). The BMD is based on age; where the elder the person, the lower the BMD. This ratio is called (T-score). If the T-score is equal to or less than -2.5, the osteoporosis is clinically diagnosed [1]. Not only the BMD, but also the peak bone mass (PBM) of early adulthood is considered as a clinical marker for osteoporosis [2].

There are two reasons describe why women are more susceptible for osteoporosis incidences. Menopause is the first cause of osteoporosis in women, due to estrogen deficiency and in turn bone resorption compared to bone formation [17, 32]. The second one starts after 4-8 years of the first phase peaks, causing a persistent bone resorption, where slower loss of trabecular and cortical bone and a decreased bone formation was observed subsequently [33].

Secondary osteoporosis can be found in both young and old individuals as a result of other diseases or a side-effect of certain medication. Hypercortisolism is the main cause of secondary osteoporosis, due to medication or Cushing's disease.

Other disorders that may cause osteoporosis are diabetes mellitus, depression, gastrectomy, hyperparathyroidism, anorexia nervosa, thyrotoxicosis, cystic fibrosis, and stroke [34].

3.1. Clinical Diagnosis of Osteoporosis

Measurement of BMD is a key marker of detection the fracture stages using related techniques [32]. The standard technique for BMD evaluation is dual-energy X-ray absorptiometry (DEXA). DEXA is a precise and an accurate technique for clinical diagnosis of lumbar spine or proximal femur by taking informative data about bone fracture and fragile regions. Usually, the BMD measurements are compared with two references; the first reference is for young adults giving a T-score and the second reference is for age matched adults giving a Z-score[1, 32]. Additionally, there is a current research interest of quantitative ultrasonography, due to its diagnostic ability for osteoporosis [1].

3.2. Current Therapy for Osteoporosis

Targeting the resorption phase by different interventions is considered to be the available therapeutic interventions for osteoporosis [5]. The anabolic interventions decrease bone resorption and remodeling. Decreasing the bone resorption by these anabolic therapeutic interventions leads to reducing the remodeling space and in turn reducing the resorption sites. This process leads to elevating bone mineralization and improving bone dimensions, thickness, and strength [35].

3.2.1. Mechanisms Behind the Current Therapy for Osteoporosis

Wu *et al.* [36] suggested that skeletal remodeling requires recruitment of osteoblast precursors, in the MSCs form, to the bone surface, mediating by osteoclastic mobilization of active TGFβ1. Regardless of osteoporosis cause, the imbalance of osteoclasts and osteoblasts is resulted, wherein the net activities of the former supersede the latter. Thus, osteoporosis therapy involves either resorption suppression or bone formation stimulation. Administration of parathyroid hormone (PTH) activates bone formation yielded the first and only bone anabolic drug [37], but the most common approach for preventing and treating osteoporosis is osteoclast inhibition.

The recruitment of either osteoclasts or osteoblasts depends upon the other. Osteoclasts differentiate from macrophage precursors under the aegis of RANK ligand and M-CSF, produced by osteoblast lineage cells. On the other side, it is the bone-resorbing activity of osteoclasts that attracts osteoblasts to the bone surface, a process known as skeletal remodeling. This ever-occurring event consists of focal removal and subsequent replacement of bone. Therefore, bone remodeling is characterized by tethering of osteoclast and osteoblast recruitment and function. Patients lose bone by two general mechanisms that are dictated by remodeling kinetics. Estrogen deficiency, a major cause of osteoporosis, is a high turnover event in which bone resorption and formation are increased, however the osteoclast activity is higher than that of the osteoblast. For decades, suppression of the osteoclast by hormone replacement had been the standard care. Due to estrogens have the risk ability to cause breast cancer and cardiovascular obstacles in old women, the osteoclasts-targeted drugs became the most common therapeutic approach for postmenopausal osteoporosis. These drugs are nonhydrolyzable pyrophosphate analogs that bind bone mineral with extremely high affinity and reside within the skeleton until mobilized by osteoclasts [38]. The capacity of bisphosphonates to enhance bone mass reflects the fact that it incapacitates osteoclasts and thus dampens remodeling.

The most common secondary form of osteoporosis is that stimulated by glucocorticoids, but its skeletal dynamics are distinctly different than those resulting from estrogen deprivation. Whereas bone remodeling is accelerated with menopause, it is suppressed by prolonged administration of glucocorticoids. Thus, glucocorticoid-induced osteoporosis represents a low remodeling form of the disease in which both formation and resorption are suppressed but the former is more so than the latter. Stimulation of bone formation by intermittent PTH administration represents the most effective current approach to this condition, but most patients are treated with bisphosphonates, which further suppress remodeling. Because osteogenesis is coupled to resorption in the remodeling process, alendronate's antiosteoclastic properties blunt the bone-forming capacity of simultaneously administered PTH [39]. Because the principal role of remodeling is probably to replace effete bone with new, its prolonged suppression probably compromises the skeleton's biomechanical properties and evidence indicates such is the case with alendronate [40]. Thus, understanding the specific mechanism by which bisphosphonates suppress osteoclast activity is central to developing antiresorptive strategies that spare bone formation.

Although remodeling has been known for decades to be essential for skeletal integrity, the means by which osteoclasts recruit osteoblasts to the site of resorption is among the most important yet enigmatic issues regarding skeletal biology and treating osteoporosis. Recently, Tang *et al*. [41] demonstrated that active TGF-β1 is a likely molecule attracting osteoblasts to sites of prior osteoclast activity. In this scenario, the activated growth factor, mobilized from

bone matrix by osteoclasts, establishes a gradient that attracts Sca1+CD29+ CD45 CD11b mesenchymal stem cells to sites of bone resorption where they putatively undergo osteoblast differentiation. To determine the mechanism by which alendronate blunts bone formation, Wu *et al*. [36] first asked whether the drug arrests PTH-stimulated osteoblast differentiation. This model was not supported by their results, despite observing reduced osteogenesis and a decreased number of bone-forming cells, *in vivo*, a circumstance reminiscent of absence of active TGF-β 1, which recruits osteoblast precursors to the bone surface [41]. In fact, the authors found that although the abundance of total TGF-β1 in bone marrow is unaffected by treatment with PTH and alendronate, alone or in combination, the hormone increases the active state of the growth factor while the bisphosphonate reduces it. This outcome is mirrored by the number of cells expressing Smad2/3, which mediates migration of osteogenic precursors to remodeling sites [41]. The authors also found that absence of TGF-β 1, in knockout mice, prevents PTH-stimulated bone formation.

Alendronate does not alter differentiation of TGF-β 1 osteoblasts, *in vitro*, but reduces their PTH-stimulated abundance, *in vivo*. Thus, the bisphosphonate, whose sole established direct target *in vivo* is osteoclasts, probably inhibits bone formation by arresting the active TGF-β1 mobilization from bone matrix, thereby preventing recruitment of mesenchymal stem cells to remodeling sites. Expanding upon their previous observations, the authors propose that the target osteoblast precursors cells are Sca1+CD29+CD45 CD11b. To determine whether such is the case, they isolated and expanded multipotential mesenchymal cells with this phenotype. Confirming their osteogenic capacity, *in vivo*, these cells form bone when placed under the renal capsule. The authors demonstrated that TGF-β1 knockout mice failed to normally recruit Sca1+ CD29+CD45 CD11b cells to the bone surface in response to PTH. These findings indicated that alendronate suppressed PTH-stimulated bone formation by inhibiting osteoclast-mediated mobilization of active TGF-β1, thereby reducing migration of osteoblast precursors to sites of remodeling. Currently, osteoporosis patients are limited to 2 year cycles of PTH treatment. Thus, a logical strategy to prevent loss of accumulated bone is to utilize an antiresorptive agent between episodes of PTH administration [39]. Wu *et al.* [36] predicted that unlike simultaneous administration of the two drugs, this sequential strategy will not suppress bone formation because osteoclast mediated mobilization of TGF-β1 will be intact during PTH-only cycles [41]. This hypothesis is challenged, however, by the decade-long persistence of bisphosphonates in the skeleton after cessation of their administration [42]. The elegant experiments of the Cao laboratory underscore the necessity of developing shortacting antiresorptive agents that will permit recovery of osteoclast-mediated mobilization of the growth factor during the period of anabolic therapy [36, 41]. The translation of observations in mice into current antiosteoporosis drugs holds promise that the present studies will yield the same [43, 44].

4. TOOTH ENGINEERING

There are potential clinical studies in engineering a whole tooth [45 - 47]. However, there are some challenges facing the ability of tooth shape and size controlling [48], ranging from tissue recombination and scaffold-based engineering approaches to gene-manipulation and chimeric tooth engineering [45 - 47, 49 - 55]. Further investigations need to be performed focusing on dental progenitor stem cells and to assess whether this novel approaches can be successfully translated into clinical world.

Nowadays, a lot of studies have investigated that there are many progenitor stem cell niches were found in the post-natal pulp, which in turn play a crucial roles in dentine formation and in wound healing, where they are also found in all body connective tissues and have high differentiation ability in response to specific stimuli. They also have a multi-potentiality property; where they can produce differentiated cells such as osteoblasts, odontoblasts, chondrocytes (Fig. **3**) [56].

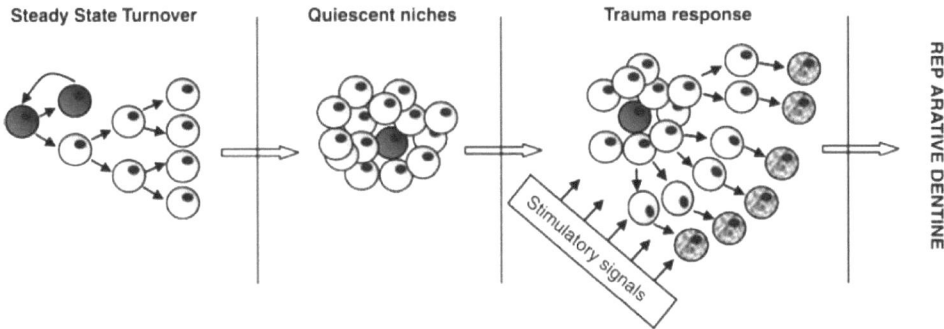

Fig. (3). The stem cell niche consists of the 'true' adult cell surrounded by the transit progenitor cells. The 'true' cell has the potentiality for infrequent self-renewal, while the daughter cells have a higher ability of proliferation along many cell lineages in response to specific stimuli [56].

4.1. Stem Cells Therapy and Dentine Regeneration

Injury and trauma are the stimuli that trigger the proliferation and differentiation activities of the progenitor stem cells. The progenitor stem cell niches were *in situ* observed after their enhancement by specific signals. Studies investigated that the replacement of odontoblasts derived from MSCs play a crucial role in synthesizing a reparative dentine [57, 58]. Moreover, cell niches were studied *in vivo* by Notch expression increment followed by pulpal injury and this experiment added further niches to the perivascular ones [59].

There is an ongoing interest in converting the current research approaches of stem cells identification and characterization within the dental regions into applied clinical strategies. The development of such novel approaches will not be easy to be applied clinically, due to the complexity of the injured pulp regeneration protocols [60]. It is well known that any inflammation in the dental regions is associated with recruitment of tooth progenitor cells to stimulate the repairing mechanisms, where those cells migrate to injury regions from other niches [61]. These migrated cells differentiate into odontoblasts to be considered as a natural reparative mechanism. Furthermore, scientists suggested that the formation of dentine bridge is triggered by calcium hydroxide [62]. Eventually, the great interest nowadays of dental regeneration processes will provides novel clinical approaches to be applied in the near future [63, 64].

4.2. Dental Pulp Stem Cells and Aging

Histomorphometric assay facilitates the assessment of pulpal cells quantification, which is reduced usually in older people [65, 66]. It was found clinically that pulpal wound healing and cell regeneration may be compromised with increasing age, where it was observed that approximately 1% potential stem progenitor cells of the total cells in the dental pulp. This small flow of stem is associated with a relatively few cells are able to help in tissue regeneration process and in turn in the stem cells therapeutic approaches [67]. Due to this defect in cells recruitment, progenitor cells cryopreservation of dental pulp was established for clinical use and it is now commercially available [68].

CONCLUSION

A fundamental understanding of the vital role of stem cells in management of bone diseases was illustrated in the current book chapter. This strategy of stem cell regeneration may provide novel clinical interventions for bone disorders. Applying such this novel strategy must be associated with fully understanding of the stem cell behavior, machineries, and networks, as well as its crucial role in clinical modalities for tackling bone diseases.

CONFLICT OF INTEREST

The author confirms that author has no conflict of interest to declare for this publication.

ACKNOWLEDGEMENTS

None Declared.

REFERENCES

[1]   Eastell R. Identification and management of osteoporosis in older adults. Medicine 2013; 41: 47-52.
       [http://dx.doi.org/10.1016/j.mpmed.2012.10.007]

[2]   Eleftheriou KI, Rawal JS, James LE, *et al.* Bone structure and geometry in young men: the influence of smoking, alcohol intake and physical activity. Bone 2013; 52(1): 17-26.

[http://dx.doi.org/10.1016/j.bone.2012.09.003] [PMID: 22985892]

[3] Rachner TD, Khosla S, Hofbauer LC. Osteoporosis: now and the future. Lancet 2011; 377(9773): 1276-87.
[http://dx.doi.org/10.1016/S0140-6736(10)62349-5] [PMID: 21450337]

[4] Ersoy FF. Osteoporosis in the elderly with chronic kidney disease. Int Urol Nephrol 2007; 39(1): 321-31.
[http://dx.doi.org/10.1007/s11255-006-9109-2] [PMID: 17103030]

[5] Luhmann T, Germershaus O, Groll J, Meinel L. Bone targeting for the treatment of osteoporosis. J Control Release 2012; 161(2): 198-213.
[http://dx.doi.org/10.1016/j.jconrel.2011.10.001] [PMID: 22016072]

[6] Tang Y, Tang W, Lin Y, *et al.* Combination of bone tissue engineering and BMP-2 gene transfection promotes bone healing in osteoporotic rats. Cell Biol Int 2008; 32(9): 1150-7.
[http://dx.doi.org/10.1016/j.cellbi.2008.06.005] [PMID: 18638562]

[7] Lane NE, Silverman SL. Anabolic therapies. Curr Osteoporos Rep 2010; 8(1): 23-7.
[http://dx.doi.org/10.1007/s11914-010-0005-4] [PMID: 20425087]

[8] Quarto R, Mastrogiacomo M, Cancedda R, *et al.* Repair of large bone defects with the use of autologous bone marrow stromal cells. N Engl J Med 2001; 344(5): 385-6.
[http://dx.doi.org/10.1056/NEJM200102013440516] [PMID: 11195802]

[9] Horwitz EM, Gordon PL, Koo WK, *et al.* Isolated allogeneic bone marrow-derived mesenchymal cells engraft and stimulate growth in children with osteogenesis imperfecta: Implications for cell therapy of bone. Proc Natl Acad Sci USA 2002; 99(13): 8932-7.
[http://dx.doi.org/10.1073/pnas.132252399] [PMID: 12084934]

[10] Arana-Chavez VE, Massa LF. Odontoblasts: the cells forming and maintaining dentine. Int J Biochem Cell Biol 2004; 36(8): 1367-73.
[http://dx.doi.org/10.1016/j.biocel.2004.01.006] [PMID: 15147714]

[11] Smith AJ, Cassidy N, Perry H, Bègue-Kirn C, Ruch J-V, Lesot H. Reactionary dentinogenesis. Int J Dev Biol 1995; 39(1): 273-80.
[PMID: 7626417]

[12] Weatherholt AM, Fuchs RK, Warden SJ. Specialized connective tissue: bone, the structural framework of the upper extremity. J Hand Ther 2012; 25(2): 123-31.
[http://dx.doi.org/10.1016/j.jht.2011.08.003] [PMID: 22047807]

[13] de Baat P, Heijboer MP, de Baat C. Development, physiology, and cell activity of bone. Ned Tijdschr Tandheelkd 2005; 112(7): 258-63.
[PMID: 16047964]

[14] Bayliss L, Mahoney DJ, Monk P. Normal bone physiology, remodelling and its hormonal regulation. Surgery 2012; 30: 47-53.

[15] Marieb EN. Human Anatomy & Physiology. 4th ed., Menlo Park, California: Benjamin/Cummings Science Publishing 1998.

[16] Kartsogiannis V, Ng KW. Cell lines and primary cell cultures in the study of bone cell biology. Mol Cell Endocrinol 2004; 228(1-2): 79-102.

[http://dx.doi.org/10.1016/j.mce.2003.06.002] [PMID: 15541574]

[17]   Kular J, Tickner J, Chim SM, Xu J. An overview of the regulation of bone remodelling at the cellular level. Clin Biochem 2012; 45(12): 863-73.
[http://dx.doi.org/10.1016/j.clinbiochem.2012.03.021] [PMID: 22465238]

[18]   Lips P, Courpron P, Meunier PJ. Mean wall thickness of trabecular bone packets in the human iliac crest: changes with age. Calcif Tissue Res 1978; 26(1): 13-7.
[http://dx.doi.org/10.1007/BF02013227] [PMID: 737547]

[19]   Tolar J, Teitelbaum SL, Orchard PJ. Osteopetrosis. N Engl J Med 2004; 351(27): 2839-49.
[http://dx.doi.org/10.1056/NEJMra040952] [PMID: 15625335]

[20]   Gillespie MT. Impact of cytokines and T lymphocytes upon osteoclast differentiation and function. Arthritis Res Ther 2007; 9(2): 103.
[http://dx.doi.org/10.1186/ar2141] [PMID: 17381830]

[21]   Horowitz MC, Lorenzo JA. B lymphocytes and the skeleton. Ann N Y Acad Sci 2007; 1117: 82-93.
[http://dx.doi.org/10.1196/annals.1402.045] [PMID: 17872391]

[22]   Elefteriou F. Regulation of bone remodeling by the central and peripheral nervous system. Arch Biochem Biophys 2008; 473(2): 231-6.
[http://dx.doi.org/10.1016/j.abb.2008.03.016] [PMID: 18410742]

[23]   Raggatt LJ, Partridge NC. Cellular and molecular mechanisms of bone remodeling. J Biol Chem 2010; 285(33): 25103-8.
[http://dx.doi.org/10.1074/jbc.R109.041087] [PMID: 20501658]

[24]   Dougall WC, Glaccum M, Charrier K, *et al.* RANK is essential for osteoclast and lymph node development. Genes Dev 1999; 13(18): 2412-24.
[http://dx.doi.org/10.1101/gad.13.18.2412] [PMID: 10500098]

[25]   Palmqvist P, Persson E, Conaway HH, Lerner UH. IL-6, leukemia inhibitory factor, and oncostatin M stimulate bone resorption and regulate the expression of receptor activator of NF-kappa B ligand, osteoprotegerin, and receptor activator of NF-kappa B in mouse calvariae. J Immunol 2002; 169(6): 3353-62.
[http://dx.doi.org/10.4049/jimmunol.169.6.3353] [PMID: 12218157]

[26]   Lee SH, Rho J, Jeong D, *et al.* v-ATPase V0 subunit d2-deficient mice exhibit impaired osteoclast fusion and increased bone formation. Nat Med 2006; 12(12): 1403-9.
[http://dx.doi.org/10.1038/nm1514] [PMID: 17128270]

[27]   Arai F, Miyamoto T, Ohneda O, *et al.* Commitment and differentiation of osteoclast precursor cells by the sequential expression of c-Fms and receptor activator of nuclear factor kappaB (RANK) receptors. J Exp Med 1999; 190(12): 1741-54.
[http://dx.doi.org/10.1084/jem.190.12.1741] [PMID: 10601350]

[28]   Parfitt AM. Targeted and nontargeted bone remodeling: relationship to basic multicellular unit origination and progression. Bone 2002; 30(1): 5-7.
[http://dx.doi.org/10.1016/S8756-3282(01)00642-1] [PMID: 11792557]

[29]   Tatsumi S, Ishii K, Amizuka N, *et al.* Targeted ablation of osteocytes induces osteoporosis with defective mechanotransduction. Cell Metab 2007; 5(6): 464-75.

[http://dx.doi.org/10.1016/j.cmet.2007.05.001] [PMID: 17550781]

[30]   Brown JP, Albert C, Nassar BA, *et al.* Bone turnover markers in the management of postmenopausal osteoporosis. Clin Biochem 2009; 42(10-11): 929-42.
[http://dx.doi.org/10.1016/j.clinbiochem.2009.04.001] [PMID: 19362543]

[31]   Garnero P, Delmas PD. Bone Turnover Markers Encyclopedia of Endocrine Diseases. 2004; 1: pp. 401-13.
[http://dx.doi.org/10.1016/B0-12-475570-4/00218-3]

[32]   Peel N. Disorders of bone metabolism. Surgery 2012; 30: 61-6.

[33]   Riggs BL, Khosla S, Melton LJ III. Sex steroids and the construction and conservation of the adult skeleton. Endocr Rev 2002; 23(3): 279-302.
[http://dx.doi.org/10.1210/edrv.23.3.0465] [PMID: 12050121]

[34]   Legrand E, Chappard D, Pascaretti C, *et al.* Trabecular bone microarchitecture, bone mineral density, and vertebral fractures in male osteoporosis. J Bone Miner Res 2000; 15(1): 13-9.
[http://dx.doi.org/10.1359/jbmr.2000.15.1.13] [PMID: 10646109]

[35]   Brixen K, Abrahamsen B, Kassem M. Prevention and treatment of osteoporosis in women. Curr Obstet Gynaecol 2005; 15: 251-8.
[http://dx.doi.org/10.1016/j.curobgyn.2005.05.003]

[36]   Wu X, Pang L, Lei W, *et al.* Inhibition of Sca-1-positive skeletal stem cell recruitment by alendronate blunts the anabolic effects of parathyroid hormone on bone remodeling. Cell Stem Cell 2010; 7(5): 571-80.
[http://dx.doi.org/10.1016/j.stem.2010.09.012] [PMID: 21040899]

[37]   Neer RM, Arnaud CD, Zanchetta JR, *et al.* Effect of parathyroid hormone (1-34) on fractures and bone mineral density in postmenopausal women with osteoporosis. N Engl J Med 2001; 344(19): 1434-41.
[http://dx.doi.org/10.1056/NEJM200105103441904] [PMID: 11346808]

[38]   Bone HG, Hosking D, Devogelaer JP, *et al.* Ten years' experience with alendronate for osteoporosis in postmenopausal women. N Engl J Med 2004; 350(12): 1189-99.
[http://dx.doi.org/10.1056/NEJMoa030897] [PMID: 15028823]

[39]   Black DM, Greenspan SL, Ensrud KE, *et al.* The effects of parathyroid hormone and alendronate alone or in combination in postmenopausal osteoporosis. N Engl J Med 2003; 349(13): 1207-15.
[http://dx.doi.org/10.1056/NEJMoa031975] [PMID: 14500804]

[40]   Allen MR, Burr DB. Three years of alendronate treatment results in similar levels of vertebral microdamage as after one year of treatment. J Bone Miner Res 2007; 22(11): 1759-65.
[http://dx.doi.org/10.1359/jbmr.070720] [PMID: 17663638]

[41]   Tang Y, Wu X, Lei W, *et al.* TGF-beta1-induced migration of bone mesenchymal stem cells couples bone resorption with formation. Nat Med 2009; 15(7): 757-65.
[http://dx.doi.org/10.1038/nm.1979] [PMID: 19584867]

[42]   Drake MT, Clarke BL, Khosla S. Bisphosphonates: mechanism of action and role in clinical practice. Mayo Clin Proc 2008; 83(9): 1032-45.
[http://dx.doi.org/10.4065/83.9.1032] [PMID: 18775204]

[43] Eisman JA, Bone HG, Hosking DJ, *et al.* Odanacatib in the treatment of postmenopausal women with low bone mineral density: three-year continued therapy and resolution of effect. J Bone Miner Res 2011; 26(2): 242-51.
[http://dx.doi.org/10.1002/jbmr.212] [PMID: 20740685]

[44] McClung MR, Lewiecki EM, Cohen SB, *et al.* AMG 162 Bone Loss Study Group. Denosumab in postmenopausal women with low bone mineral density. N Engl J Med 2006; 354(8): 821-31.
[http://dx.doi.org/10.1056/NEJMoa044459] [PMID: 16495394]

[45] Duailibi MT, Duailibi SE, Young CS, Bartlett JD, Vacanti JP, Yelick PC. Bioengineered teeth from cultured rat tooth bud cells. J Dent Res 2004; 83(7): 523-8.
[http://dx.doi.org/10.1177/154405910408300703] [PMID: 15218040]

[46] Ohazama A, Modino SA, Miletich I, Sharpe PT. Stem-cell-based tissue engineering of murine teeth. J Dent Res 2004; 83(7): 518-22.
[http://dx.doi.org/10.1177/154405910408300702] [PMID: 15218039]

[47] Duailibi SE, Duailibi MT, Zhang W, Asrican R, Vacanti JP, Yelick PC. Bioengineered dental tissues grown in the rat jaw. J Dent Res 2008; 87(8): 745-50.
[http://dx.doi.org/10.1177/154405910808700811] [PMID: 18650546]

[48] Yu J, Shi J, Jin Y. Current approaches and challenges in making a bio-tooth. Tissue Eng Part B Rev 2008; 14(3): 307-19.
[http://dx.doi.org/10.1089/ten.teb.2008.0165] [PMID: 18665759]

[49] Young CS, Terada S, Vacanti JP, Honda M, Bartlett JD, Yelick PC. Tissue engineering of complex tooth structures on biodegradable polymer scaffolds. J Dent Res 2002; 81(10): 695-700.
[http://dx.doi.org/10.1177/154405910208101008] [PMID: 12351668]

[50] Yu J, Deng Z, Shi J, *et al.* Differentiation of dental pulp stem cells into regular-shaped dentin-pulp complex induced by tooth germ cell conditioned medium. Tissue Eng 2006; 12(11): 3097-105.
[http://dx.doi.org/10.1089/ten.2006.12.3097] [PMID: 17518625]

[51] Young CS, Abukawa H, Asrican R, *et al.* Tissue-engineered hybrid tooth and bone. Tissue Eng 2005; 11(9-10): 1599-610.
[http://dx.doi.org/10.1089/ten.2005.11.1599] [PMID: 16259613]

[52] Yu JH, Shi JN, Deng ZH, *et al.* Cell pellets from dental papillae can reexhibit dental morphogenesis and dentinogenesis. Biochem Biophys Res Commun 2006; 346(1): 116-24.
[http://dx.doi.org/10.1016/j.bbrc.2006.05.096] [PMID: 16750168]

[53] Edwards PC, Mason JM. Gene-enhanced tissue engineering for dental hard tissue regeneration: (1) overview and practical considerations. Head Face Med 2006; 2: 12.
[http://dx.doi.org/10.1186/1746-160X-2-12] [PMID: 16700908]

[54] Edwards PC, Mason JM. Gene-enhanced tissue engineering for dental hard tissue regeneration: (2) dentin-pulp and periodontal regeneration. Head Face Med 2006; 2: 16.
[http://dx.doi.org/10.1186/1746-160X-2-16] [PMID: 16725030]

[55] Cordeiro MM, Dong Z, Kaneko T, *et al.* Dental pulp tissue engineering with stem cells from exfoliated deciduous teeth. J Endod 2008; 34(8): 962-9.
[http://dx.doi.org/10.1016/j.joen.2008.04.009] [PMID: 18634928]

[56]   Sloan AJ, Waddington RJ. Dental pulp stem cells: what, where, how? Int J Paediatr Dent 2009; 19(1): 61-70.
[http://dx.doi.org/10.1111/j.1365-263X.2008.00964.x] [PMID: 19120509]

[57]   Fitzgerald M, Chiego DJ Jr, Heys DR. Autoradiographic analysis of odontoblast replacement following pulp exposure in primate teeth. Arch Oral Biol 1990; 35(9): 707-15.
[http://dx.doi.org/10.1016/0003-9969(90)90093-P] [PMID: 2091590]

[58]   Tziafas D, Kolokuris I. Inductive influences of demineralized dentin and bone matrix on pulp cells: an approach of secondary dentinogenesis. J Dent Res 1990; 69(1): 75-81.
[http://dx.doi.org/10.1177/00220345900690011301] [PMID: 2303600]

[59]   Løvschall H, Tummers M, Thesleff I, Füchtbauer EM, Poulsen K. Activation of the Notch signaling pathway in response to pulp capping of rat molars. Eur J Oral Sci 2005; 113(4): 312-7.
[http://dx.doi.org/10.1111/j.1600-0722.2005.00221.x] [PMID: 16048523]

[60]   Rutherford RB, Gu K. Treatment of inflamed ferret dental pulps with recombinant bone morphogenetic protein-7. Eur J Oral Sci 2000; 108(3): 202-6.
[http://dx.doi.org/10.1034/j.1600-0722.2000.108003202.x] [PMID: 10872990]

[61]   Téclès O, Laurent P, Zygouritsas S, *et al.* Activation of human dental pulp progenitor/stem cells in response to odontoblast injury. Arch Oral Biol 2005; 50(2): 103-8.
[http://dx.doi.org/10.1016/j.archoralbio.2004.11.009] [PMID: 15721135]

[62]   Graham L, Cooper PR, Cassidy N, Nor JE, Sloan AJ, Smith AJ. The effect of calcium hydroxide on solubilisation of bio-active dentine matrix components. Biomaterials 2006; 27(14): 2865-73.
[http://dx.doi.org/10.1016/j.biomaterials.2005.12.020] [PMID: 16427123]

[63]   Nör JE. Tooth regeneration in operative dentistry. Oper Dent 2006; 31(6): 633-42.
[http://dx.doi.org/10.2341/06-000] [PMID: 17153970]

[64]   Sloan AJ, Smith AJ. Stem cells and the dental pulp: potential roles in dentine regeneration and repair. Oral Dis 2007; 13(2): 151-7.
[http://dx.doi.org/10.1111/j.1601-0825.2006.01346.x] [PMID: 17305615]

[65]   Murray PE, Stanley HR, Matthews JB, Sloan AJ, Smith AJ. Ageing human odontometric analysis. Oral Surg Oral Med Oral Pathol Oral Radiol 2002; 93: 474-82. a
[http://dx.doi.org/10.1067/moe.2002.120974]

[66]   Murray PE, Matthews JB, Sloan AJ, Smith AJ. Analysis of incisor pulp cell populations in Wistar rats of different ages. Arch Oral Biol 2002; 47(10): 709-15. b
[http://dx.doi.org/10.1016/S0003-9969(02)00055-9] [PMID: 12356502]

[67]   Smith AJ, Patel M, Graham L, Sloan AJ, Cooper PR. molecular signalling. Oral Biosci Med 2005; 2: 127-32.

[68]   Perry BC, Zhou D, Wu X, *et al.* Collection, cryopreservation, and characterization of human dental pulp-derived mesenchymal stem cells for banking and clinical use. Tissue Eng Part C Methods 2008; 14(2): 149-56.
[http://dx.doi.org/10.1089/ten.tec.2008.0031] [PMID: 18489245]

CHAPTER 11

A Scope on Stem Cells and Human Parasites

Marwa Adel HasbySaad[*]

Faculty of Medicine, Tanta University, Tanta, Egypt

Abstract: In this chapter, the aim is to shed the light if the new era of stem cells can play a role in the field of Medical Parasitology. We will try to answer certain questions. First; would parasites be friends or foes in the process of stem cell culture and therapy? Another question is that; can stem cells be a novel therapy against the notorious parasites that attack the human being and heal the permanent damage that some parasites may induce in organs? Finally, could the parasite stem cells be a potential target for new anti-parasitic therapy, especially in resistant chronic debilitating parasitism?

Keywords: ADMSCs, Alveolar ecchinochoccosis, Chronic Chagas Cradiomyopathy, Cryptosporidium, HPSCs, Malaria, Leishmaniasis, Neoblast-like stem cells, Parasites, Schistosomiasis, Toxoplasmosis.

1. MEDICAL PARASITOLOGY; *WHAT DOES IT MEAN?*

Although Parasitology had been originally a zoological science, the more précising term "Medical parasitology" would refer to the study of a group of pathogens classified as eukaryotes, kingdom Animalia, that cause a wide range of diseases in the human beings. Parasites infecting the humans can be generally divided into three broad categories; parasitic protozoa, parasitic helminths (worms), and arthropods. While the first two categories affect humans by tissue invasion and irritation, the third one can cause diseases either directly by tissue invasion, envenomization , allergic manifestations after biting or by acting as

---

[*] **Correspondence author Marwa Adel HasbySaad:** Faculty of Medicine, Tanta University, Tanta, Egypt; E-mails: m.hasby@yahoo.com, marwa.saad@med.tanta.edu.eg

Ahmed El-Hashash (Ed)

vectors or vehicles that transmit various infectious agents to the humans, bio-logically and mechanically [1]. Generally, parasitic diseases fall under the broader term "infectious diseases" which include bacterial, viral, mycotic and parasitic diseases. They are considered the second most common cause of deaths all over the world after cardiovascular diseases, proceeding malignant neoplasms which come in the third grade [2]. Malaria which is caused by the parasitic protozoan *"Plasmodium"* ranks early in the list [3]. The human infections inflicted by parasites are numbered in billions. They vary between relatively innocuous infections to fatal diseases, especially with the ability of some of them to spread all over the body systems. The diseases caused by parasites are considered a worldwide major health problem. For example, *Ascaris lumbricoides* infection has reached 30% of the world's population [4]. Also, the incidence of schistosomiasis and malaria has been increasing rather than decreasing. Some parasitic illnesses are known to be opportunistic (*e.g.*, cryptosporidiosis, toxoplasmosis, and strongyloidiasis) [5]. The hazards of opportunistic parasites have drawn much attention in the last century due to the epidemics of AIDS, and immune suppression after radiotherapy and immunosuppressive drugs [6]. Unfortunately, parasites have a wide range of modes and routes for transmission. They can reach humans in civilized areas due to pollution in the form of contaminated drinking water, improper sewage disposal (*e.g. Cryptosporidium* oocysts, *Entamoeba histolytica* and *Giardia lamblia* cysts), bad hygiene, exposure to soil contaminated with the helminth larvae and eggs (*e.g.* Hook worms & *Trichuris trichuira*) or after exposure to insects' bites like Mosquitoes that transmit malaria and *Tse tse* fly that transmit sleeping sickness. Parasites can get access to our bodies through almost all the routes of transmission; oral, inhalation, contact, skin penetration either passive or active, congenital, blood transfusion and by organ transplantation [7, 8]. Though parasites are famous for being endemic in the neglected tropical areas, lately, many people from the temperate and subtropical areas have become also infected with parasites. This can be explained by the phenomenon of global warming and the changes in climate conditions, which, for example, enhance the parasites to continue its life cycle. Migration from or visiting tropical countries also consist an important factor in increasing the spread of parasitic diseases [7, 9]. Parasites vary greatly during their life cycle. Some helminths have larval stages that little resemble the adult stage (for example, tapeworms and flukes).

Also parasites may undergo many changes during their life cycle. They may be transmitted to humans from animals (zoonotic infections) but with a completely different clinical picture. When *Toxoplasma gondii*, which is an intestinal coccidian protozoan in cats, infects the human beings it takes a different form and become localizes in deep tissues [1]. The broad spectrum of parasites, starting from unicellular parasites (protozoa) to multicellular parasites (helminths, arthropods) does not show only changes in morphology but also in biochemical, antigenic and genetic library during their life aspects [10]. This makes the battle against parasites in prevention, control and treatment not an easy task. Hence, parasites are common to have a chronic course causing debilitation and permanent tissue damage with common development of drug resistance. There is a tremendous interplaying between the parasite and the human immune system. It starts from invasion of the first defense line in the form of the skin and mucous membrane till reacting with, escaping, tuning or even hi-jacking the highly sophisticated cells and cytokines of our immune system. This made parasites suitable organisms in understanding the immune reactions in the human body and a rich environment for medical biochemical, genetic and immunological researches.

2. PARASITIC INFECTION: WOULD IT BE A HAZARD AFTER STEM CELL TRANSPLANTATION?

Protozoa are one-celled parasites, with a worldwide prevalence. In humans, infections range from asymptomatic to fatal diseases, depending on the protozoan species and strain and the host resistance. The unicellular character of protozoa with the ability of some species to invade human cells makes some protozoa easily transmitted during blood transfusion and organ transplantation. Five % of more than 340 known infectious diseases reported after transplantation have been caused by protozoa, either due to a transmitted infection from the donor or the relapse of a dormant infection in the recipient after immune suppression.

The fact that several protozoa (*e.g. Toxoplasma, Cryptosporidium,* and *Plasmodium* species) could create a risk factor in stem cell transplantations has been reported in several studies.

2.A. Relapse of a Dormant Infection

2.A.1. Cryptosporidiosis

Recently, the world has witnessed an obvious increase in bone marrow stem cell transplantation for the treatment for leukemia and other hematological malignancies. There has been a growing fear of the risk that these patients may catch an infection with the opportunistic protozoan "*Cryptosporidium*" after bone marrow transplantation [11]. *Cryptosporidium* is a parasitic protozoan that is present in the form of oocyst. There are many species of *Cryptosporidium*, which infect humans and animals to habitat the brush border of the small intestine. In the external environment, the parasite is protected by an outer shell in the form of a thick walled oocyst to allow survival for a long time outside the body [12]. Water is the most common route of transmission, beside several different ways including faeco-oral transmission and contact with animals. The microscopic size of this protozoan (4-6 μ) makes it easily transmitted and inhaled through the air. Watery diarrhea and abdominal cramps are often the main complaints. Other symptoms may include nausea, vomiting and weight loss. Symptoms are either continuous or remittent and usually occur a week after exposure [13]. *Cryptosporidium* is well known for causing severe life-threatening manifestations in immunocompromised patients like in AIDS. The severity of AIDS is related to the CD4 count. *Cryptosporidium* has been also reported to cause gastritis, sclerosing cholangitis, pancreatitis and pneumonitis in patients with lukaemia and receiving chemotherapy, in addition to harsh diarrhea in during or after bone marrow transplantation [14]. There have been several reports of [15]. An Italian paper reported the progress of cryptosporidiosis in patients, suffering originally from a hematological malignancy. Most of the patients were adults, who conducted bone marrow transplantation. Of the twenty patients, 25% had severe diarrhea, 50% had moderate diarrhea, and the rest 25% were asymptomatic carriers. One patient had extraintestinal infection in the form of pulmonary involvement and finally died from such intractable cryptosporidiosis. After recovery of the remaining patients, four of them suffered from cryptosporidial relapses. Gentile *et al*. also represented two typical cases of copious diarrhea [16], as Manivel *et al*. The patients in the two papers experienced pulmonary cryptosporidiosis. Two patients died, while one patient managed to survive after treatment [17].

There has been an outbreak of cryptosporidiosis in a bone marrow transplantation unit [18]. Cryptosporidiosis has been also noticed after T cell depletion and severe immunosuppression due to graft *versus* host disease (GVHD). Nachbaur and his colleagues reported two cases of cryptosporidiosis after CD34- selected PBSCT for lymphoma [49]. Unfortunately, routine stool examinations can not detect this parasite, as it needs special stains (like modified Ziel Nelseen), duodenal aspiration and biopsy and copro-antigen detection. In the first patient, infection was diagnosed after endoscopic duodenal biopsy [11]. Treatment with paromomycin showed no response, and the patient's condition deteriorated to die with respiratory failure after pulmonary cryptosporidiosis. However, treatment of the second patient with azithromycin and subcutaneous low-dose rhIL-2 in addition to paromomycin, succeed to improve T lymphocyte defects and elimination of infection. They concluded that the removal of mature T lymphocytes by positive selection of CD34 cells delays the immune reconstitution and increases the rate of severe to fatal opportunistic infections. IL-2 may be also useful as it accelerates immune reconstitution and decreases the danger of opportunistic infections [19]. In conclusion, Cryptosporidiosis a possible and may be fatal hazard after autologous bone marrow and transplantation of peripheral blood stem cell.

Although some reports describe a beneficial effect of spiramycin, paromomycin or azithromycin [20], still effective treatment against cryptosporidiosis in such cases is lacking. So, routine investigation for *Cryptosporidium* in the recipient patients for early diagnosis is a must. Also, immune reconstitution or reduction of the doses of immunosuppressive drug is prerequisite to eliminate *Cryptosporidium*.

2.A.2. Toxoplasmosis

Toxoplasmosis after haematopoietic stem cell transplantation (HSCT) is a severe infection with high fatality even with early diagnosis. It occurs in allogeneic transplant recipients mainly, and to a less extent in autologous transplant recipients. *Toxoplasma gondii* is an intracellular opportunistic protozoal parasite.

Transmission to man occurs through; consumption of undercooked meat containing tissue cysts; exposure to oocysts in soil contaminated with cat feces or transplacental transmission. Transmission may also occur through blood transfusion and solid organ transplantation. In healthy individuals, primary infection is often asymptomatic and accidentally diagnosed by serological techniques [21]. The probability of catching infection and become *Toxoplasma* seropositive increases with age due to the increase in the exposure rate. However, symptomatic infection may be clinically manifested as follows: (1) mild lymphatic form that resembles infectious mononucleosis; (2) chronic toxoplasmosis with eye affection in the form of retinochoroiditis; (3) an acute disseminated infection; or (4) single-organ involvement, like affecting the brain in the form of encephalitis. The latter two forms occur mainly in severe immune suppression and reactivation of latent infection in the seropositive patients [22]. Toxoplasmosis has been reported as the cause of death after HSCT as were diagnosed in twenty-two cases, in a study by Martino *et al.* [23]. The study represented the longest series of patients with toxoplasmosis after allogeneic and autologous HSCT. It was carried out on patients with HSCT in 15 different transplant centers in Europe from 1994 till 1998. However, among autologous recipients, only 2 patients were represented by classic lymphadenopathy after the autologous bone marrow transplantation and both patients recovered without therapy. This confirms the rarity of this complication in this setting. In the other hand, 41 cases were already diagnosed with toxoplasmosis after allogeneic HSTCs. Ninety four percent of patients were *Toxoplasma gondii* seropositive before transplantation, and 73% developed moderate to severe acute graft-*versus*-host disease before being manifested with toxoplasmosis. Eighty five percent of patients had toxoplasmosis with organ involvement, whereas 15% patients experienced fever and a positive blood samples for *T. gondii* by PCR. In previous reviews by; Derouin *et al.* Chandrasekar *et al.* and Sing *et al.* 55 patients were reported to have toxoplasmosis after HSCT. However, this is considered small number in comparison to the high frequency of such infection in advanced AIDS patients [22, 24, 25].

Since *Toxoplasma* reactivation proved to be a life-threatening danger after allogeneic stem cell transplantation, especially with the difficulty of detection of parasites in tissue during diagnosis, Costa and his group have worked on developing a real-time quantitative PCR test using fluorescence hybridization technique for the detection of *Toxoplasma gondii* DNA in serum. This PCR test has given reproducible quantitative results of a wide dynamic range. This would be very useful for the diagnosis and follow-up of *Toxoplasma* reactivation, especially after allogeneic stem cell transplantation, to start the treatment early while the parasite burden is still low. This avoids the patient the invasive diagnostic procedures such as brain biopsy [26].

2.B. Parasite Transmission Through Stem Cells Transplantation

2.B.1. Leishmaniasis

Unlike *Toxoplasma* and *Cryptosporidium*, there is only one reported case that has been published about the flaring up of *Leishmania* parasites in a patient with haemopoietic stem cell transplantation. The case was diagnosed by retrospective western blot analysis [27]. This may be due to the facts that leishmaniasis can be misdiagnosed due to its difficulty in diagnosis. Also, leishmaniasis is often neglected or underestimated from the clinical point of view [28]. *Leishmania* species are protozoal haemoflagellates that are transmitted where their vector "the Sand fly" is prevalent like in Mexico, South America, Africa and India in South Asia. After the bite of the blocked Sand fly, *Leishmania* promastigotes are engulfed with receptor- dependent phagocytosis by skin macrophages. Phagocytosed promastigotes turn into amastigote forms, proliferate, and are released by bursting the host cell. According to the species of the transmitted *Leishmania*, extend to the nearby mucocutaneous junction. and settle in reticuloendothelial system of different organs. For example, some species settle in the skin, another may extend to the nearby mucocutaneous junction. *Leishmania donovani* is the most virulent species as the amastigotes have the ability to spread into RES cells all over the body, e.g. in the spleen, liver, GIT and bone marrow in addition to the skin [29]. *Leishmania* uses sophisticated strategies to stop the normal functions of macrophage, for example; it prevents activation of nitric oxide and inhibits many cytokine mediated macrophage functions. These escape techniques enable the parasite to avoid the immune attacks and flourish intracellularly inside the phagolysosome, to spread to the other organs. However, according to Rebelatto *et al*. [30], *Leishmania* amastigote are not just professional to invade RES cells, but they are also able to invade other cell types such as amniotic epithelial cells, human epithelial cells, kidney cells, dendritic cells and fibroblast cells in mammalian hosts. The umbilical cord and adipose tissue are also considered of the richest sources of mesenchymal stem cells after bone marrow. This occupied an important role in mesenchymal stem cells research and therapeutic studies for many years. Lately, adipose tissue has become even more superior over the bone marrow, because of the easy procedure, low morbidity, and high volume of stem cells aspirated from it [31] As Allahverdiyev *et al*. [32] found no studies that report the transmission of *Leishmania* parasites through ADMSCs, they started to investigate the interaction between ADMSCs and the different species of *Leishmania* parasites *in vitro* and the infectivity status of mesenchymal stem cells by *Leishmania* parasites, especially that ADMSCs were reported to have a phagocytic property in a study done by Mazo and his group in 2008 [33]. This makes leishmaniasis a possible transmitted infection through stem cell therapy. *Leishmania* parasites can survive in different tissues and organs for decades as latent inactive forms, even after treatment. They can keep their viability not only in fibroblasts, but also in mesenchymal stem cells. Allahverdiyev *et al*. [32] proved that visceral and cutaneous *Leishmania* species succeeded to infect ADMSCs. This can cause a major problem in case of stem cell transplantations in regions endemic for leishmanisis. So, performing a screening test for such parasite is important to avoid the transmission and reactivation of *Leishmania* after taking the immunosuppressive drugs regimen, as has been reported before in several case studies on organ transplantation. The immune response responsible for host resistance or susceptibility to Leishmaniasis is T-helper (Th) 1 and Th2 dependent. Interleukin-12 driven Th1 response is widely accepted as the initial protective response against *Leishmania*. The problem is that the decrease in CD4+ T lymphocytes number in patients receiving immuno-suppressive drugs with transplantation would be inversely proportional to the parasite load and dissemination [34]. Subsequently, atypical manifestations of leishmaniasis, in addition to hepatosplenomegally, pancytopenia and weight loss, make the diagnosis of leishmaniasis a very challenging task, especially that the disease is frequently associated and masked by other opportunistic infections. This would end up with failure of the usual anti-leishmanial drugs. Twenty post-transplantation mortality cases had been diagnosed to be caused by leishmaniasis post-mortem [35].

2.B.2. Malaria

Malaria is the world's most common parasitic infection. Moreover, 40% of the world's population inhabits malaria endemic areas. Malaria is a mosquito-borne disease that is caused by different species of the blood borne protozoal parasite *"Plasmodium"*. People infected with malaria are manifested by fever, chills, and flu-like illness. They may develop severe complications and end up with death if left untreated.

Between 2000 and 2010, malaria mortality rates have decreased to 26%. During this period only 1.1 million malaria deaths were reported all over the world, primarily due to application of prevention and control measures [36]. Though malaria is biologically acquired dueto the bite of an infected *Anopheles* mosquito, it can be also transmitted by red blood cells transfusion and organ transplantation [37, 38].

Acute malaria has been also reported to be transmitted by bone marrow transplantation from malaria-infected donors. However, the route of transmission was uncertain as many transplant recipients have also received blood transfusion and all the procedures occurred in malaria endemic regions, where natural infection cannot be excluded [39]. In *Plasmodium* life cycle, man act as an intermediate host where the asexual part of its life cycle takes place in two sequential phases; first the hepatic or pre- eyrthrocytic phase that occurs once, then secondly the repeated erythocytic phase. In the erythrocytic phase; the parasites multiply within the red blood cells periodically ending in rupture of the infected RBCs to invade newly fresh ones, repeatedly. Such amplification cycles occur several times, ending in waves of fever, which arise from the escaping merozoites after RBCs rupture to infect new RBCs [40]. After the parasite invasion, the red cells undergo profound structural and morphological changes and become rigid [41]. By this; the intracellular location of the parasite in the liver and blood cells makes it protected somehow from the immune system, till infected RBCs become destructed in the spleen [36]. To avoid destruction, the

P. falciparum parasite presents adhesive proteins on the RBCs surface to enable sticking of the infected RBCs to the endothelial lining of small blood vessels in deep organs. This protects the parasite from returning to the general circulation and the spleen. Sequestered RBCs will block the microvasculature and cause many complications like placental and cerebral malaria. Subsequently, *Plasmodium falciparum* is more associated with severe complications and expected to be transmitted through peripheral blood stem cell transplantation [42].

Allogeneic bone marrow (allo-BMT) and blood stem cell transplants (allo-BSCT) are progressively applied in malaria endemic areas. Post-transplantation pancytopenia in these patients is a common complication and mainly caused by graft-*versus*-host disease (GVHD), infection with cytomegalovirus, immune attack or graft rejection. Pancytopenia due to malaria is very uncommon.

It was reported in three cases only from 1986 to 1997, following sibling allogeneic bone marrow transplant [43 - 46]. Then, in 1998, Raina *et al.* also reported a 20 years old male patient suffering from chronic myeloid leukemia from India, who continued to have post-transplant pancytopenia and splenomegaly after receiving allogeneic blood stem cell transplantation from his sister. Examination of peripheral smear showed the schizont stages of *Plasmodium vivax*. After chloroquine therapy, the pancytopenia reversed completely. However, this patient was more likely to have undetected infection before the transplant [45]. In 2012; Mejia *et al.* [42] described a female patient from Africa, infected with *Plasmodium falciparum* in association with sickle cell disease and auto-splenectomy. Sickle cell disease (SCD) is caused by a single nucleotide mutation in the β-globin gene, leading to increased polymerization of HbS and formation of sickle-shaped RBCs. This is manifested with anemia, hemolysis, and recurrent vessels occlusions. Life expectancy of patients decreases a lot, and the only cure is allogeneic stem cell transplantation [47, 48]. The case described by Mejia *et al.* had lived in the U.S. for more than two years without travelling to malaria endemic areas. The donor was asymptomatic for malaria, denied conducting any previous malarial infection and had three negative thick blood smears [42]. However, 11 days after transplantation, the recipient suffered from the characteristic typical malarial febrile episodes, which was proved to be caused by *Pf* transmission in the stem cell product after serological and PCR analysis. White blood cells began to recover in the patient with 96% donor source myeloid cells by the 19th. Still erythocytic intracellular ring forms were identified in the routine peripheral blood smear examination, with 3.5% malarial parasitemia. The timing of manifestations onset and the negative tests for PfHRP2 antigen prior to transplant were more directing to the potentiality of an exogenous source of parasitemia like RBC transfusions. However, that was excluded by the look-back investigation for the blood donors. *P. falciparum* was confirmed by real-time PCR, when the 14 days pre-transplant frozen plasma sample negative turned positive on day 11 post- transplant which matched the onset of fever in the recipient. For the donor, his blood was evaluated using PCR and PfHRP2 antigen ELISA. PCR testing was negative on day 17, till turned positive by day 40. In the other hand, antigen testing in blood continued to be positive from day 17 till 40, while stored plasma obtained 3 months before transplanttion had become antigen-positive for *Plasmodium falciparum*.

Though receiving the proper anti-malarial treatment, manifestations didn't resolve in the recipient till transplant rejection. Meanwhile, the donor received anti-malarial treatment prior to remobilization and apheresis. However, the second transplant to the same recipient had also failed, but without conducting fever or recurrence of parasitemia this time. What suggests that the donor had been the original source of infection is that; while PBSCT stem cells were collected by apheresis, processed and frozen at temperatures below −180 °C, some RBCs (including those infected with *Plasmodium falciparum*) were still preserved in this process, and transmitted to the recipient, especially that, free *Plasmodium falciparum* merozoites had been proved to succeed in surviving in similar freezing conditions, to be used later in *in vitro* culture.

Another explanation is that, viable extracellular parasites may have been present in the stem cell product and transmitted to the recipient. Finally, other hematopoeitic cells as monocytes and neutrophils, could have already ingested parasitized RBC and were still small in number in PBSCT stem cells [49]. Taken together, screening including both the multi-species PCR and PfHRP2 antigen ELISA tests must be consistent before PBSCT even from asymptomatic donor with negative blood smears [50].

3. STEM CELLS: WOULD THEY BE A POTENTIAL REPAIR IN PARASITIC INDUCED TISSUE DAMAGE?

Recently, stem cell therapy has proved to be a potential therapy in some parasitic diseases especially with permanent organ damages like schistosomiasis, malaria and American trypanosomiasis "Chagas disease". With further researches in the coming few years; the appropriate stem cell type, number of cells and route of injection will be accurately defined for such chronic parasitic diseases [51].

3.A. Schistosomiasis

Schistosoma infection is characterized by the formation of multiple granuloma around schistosome eggs in the liver at the early stage of the disease, followed by fibrosis in the chronic stage [52, 53]. Currently, praziquantel is the drug of choice and almost the only treatment for this neglected tropical disease. However, the potentiality of the development of praziquantel resistance makes it urge to discover a novel strategy to fight Schistosoma infection in humans [54] and to fix the *schistosoma* induced permanent organ damage in the form of liver fibrosis, complicated by ascitis, hemorrhoids and piles. After the previous successes of mesenchymal stem cells (MSCs) in the treatment of infectious diseases and the fibrotic insult infections induce in the organs [55, 56], MSCs have been proposed as a potential therapy for *S. mansoni* and *S. japonicum* infection. However, the efficacy and underlying mechanisms are still under investigations and only few studies have investigated them. In 2011, Xu *et al*. proved that MSC culture supernatant can prevent the activation and proliferation of macrophages stimulated by S. japonicum soluble egg antigen (SEA). Macrophages have become round, with a significant reduction in size and pseudopodia after incubation in the MSC culture supernatant mixed with SEA for 12 h, in comparison to cells cultured with SEA only [52].

From the Far East to the Middle East, where *Schistosoma mansoni* liver fibrosis is one of the most serious health problems in Egypt, with many people suffering from its mortal complications. Many patients suffer from a combined schistosomal liver fibrosis and hepatitis C cirrhosis. The problem had started in the last century, when the public health organization was actively involved in controlling schistosomiasis by tartar emetic injections. It was observed that almost 500 individuals were injected in using only 20 to 30 syringes in rotation, thereby not allowing adequate sterilization, which enhanced the HCV transmission [58]. Though liver transplantation has been considered the only effective treatment for post-schistosomal hepatic fibrosis, it still has its limitations due to the scarcity of donors and post-operative hazards. Hence, in Egypt 2012, a joint research in Tanta and Mansoura Universities was carried out by Dr. Abo Raya, to investigate the impact of bone marrow MSCs on schistosomal liver fibrosis [59]. The researcher investigated the effect of MSCs on chronic liver fibrosis induced by *S. mansoni* infection in mice. The study revealed that the bone marrow MSCs successfully migrated to the damaged liver and got engrafted into it, resulting in a significant morphometric and histopathological improvement with an increase in IL-10 in mice treated with combined MSCs and praziquantel (PZQ) in comparison to groups untreated or treated with praziquantel only. As a continuation of the research carried out by Xu *et al*. in 2011 [52], the same team carried out a similar research to that of Abo Raya [59], but *in vivo* on *S. japonicum* infected mice this time. They reported that MSCs were able to ameliorate *S. japonicum*-induced liver injury, enhanced by a combined MSCs and the regular praziquantel therapy. MSC treatment didn't only enhance the infected mice survival rate, but also reduced granuloma size, and serum concentrations of hyaluronic acid and transforming growth factor-β1. MSCs treatment also inhibited collagen deposition and decreased the collagen type 3 and α-smooth muscle actin expression in mice hepatic tissue. This provides an important evidence for MSCs applications in the treatment of Schistosoma-induced hepatic fibrosis [57].

3.B. Malaria

When the parasite causing malaria "*Plasmodium*" get transmitted to human by *Anopheles* mosquito bites, the patient suffers from anemia due to destruction of RBCs and causing bone marrow dyserythropoiesis [60, 61]. However, some hemoglobin variants were found to develop resistance to *Plasmodium* species, hence it was postulated that stem cell engineering may help in the production of erythrocytes with newly modified hemoglobin to be protective against severe malaria [62]. In 1991, Asami *et al*. reported that multipotent hemopoietic stem cells can play a significant role in the host's defense mechanisms against

Plasmodium berghei. After conducting *Plasmodium berghei* infection in mice deficient with haemopoietic stem cell, the survival time of mice was shorter than usual with a substantial increase in multipotent haemopoietic stem cells. Cells also committed into granulocytes and macrophages or erythrocytes (CFU-E) in the bone marrow and spleen in non-deficient mice. The number of CFU-S in the spleen was observed to restore if bone marrow grafting from positive to deficient mice was carried out 2 months prior to infection [63]. In the context of experimental malaria infection, the National Institute for Medical Research, scientists in the United Kingdom succeeded to identify a multipotent progenitor cell population capable of generating myeloid cells. When they transplanted the atypical progenitor cells recovered from acute malaria-infected mice to severe malaria ill mice, the cells were able to give rise to a generation of cells capable of fighting the disease, and helped the recipient mice to recover [64]. This can be explained that some parasitic infections such as malaria, schistosomiasis and leishmaniasis can induce haematopoiesis to renew cells destructed by infection and the inflammation and to replace the cells that have been activated and migrated from secondary lymphoid compartments to infected tissue sites [65 - 70]. According to Thakur *et al.* research in the National Institute of Malaria Research, New Delhi, India; MSCs have a protective effect against malaria through T regulatory cells modulation [71]. That was their conclusion when mice experimentally infected with P. berghei have showed marked migration of stem cell antigen-1+CD44+CD29+CD34− cells to the spleen and lymph nodes. Sca-1+CD44+CD29+CD34− cells are a phenotype that is consistent with mesen-chymal but not conventional stem cells. They stay in the bone marrow and attracted to the area of inflammation during tissue injury [71 - 73]. To test the pluripotency of cells accumulated in the spleen, Thakur and his colleagues isolated CD3−CD19−Sca-1+ cells and cultured them in adipocyte differentiation medium. They noticed that some cells succeeded to differentiate into adipocytes. They concluded that; the accumulated cells in the spleen in response to infection with *P. berghei* are bona fide MSCs with a potentiality to differentiate into different cell lineages. Furthermore, infusion of the isolated MSCs from infected mice into naïve ones has proved to enhance resistance against malaria in the recipient mice, with augmentation of IL-12 production and marked reduction of regulatory T cells and IL-10, which means that MSCs have immune modulatory activities. They increase the pro-inflammatory cytokines, and decrease the Treg cells and anti-inflammatory cytokines [74, 75]. Another finding regards spleen recruited MSCs in malarial infection, is that they fail to yield NO, whereas bone marrow-derived MSCs produce large amounts of NO in uninfected animals. Therefore, deployment of MSCs to the site of malaria infection represents a new protective strategy to combat malaria by modulating Treg-cell response [74]. However, malaria parasites resist such massive infiltration with MSCs and continue to grow causing the disease. This may be due to that the parasite had evaded the host's defense mechanisms already before the MSCs recruitment; especially that infusion of MSCs at late infection didn't alter either the disease progress or parasitaemia [62, 76]. Nonetheless, Thakur *et al*. results provide strong proof that MSCs recruitment can be considered one of the immune protective ways against malaria infection [71].

3.C. Chagas Disease

Chagas disease or American trypanosomiasis is a cardiomyopathic disease that results from infection with the protozoan *Trypanosoma cruzi* (*T. cruzi*). It's considered one of the major causes of cardiomyopathy in Latin America, and has many therapeutic limitations in both the acute and chronic stages. It is vector-borne and transmitted by *Triatoma* bug. *Triatoam* bug (also famous for kissing bug) usually bites an exposed area of skin such as the face, and then defecates close to the bite wound. The infective stage of *Trypanosoma cruzi* "the metacyclic tryposmastigote" enters the body when the person rubs the bug feces into a skin break or mucous membrane. The disease may also spread through blood transfusion, organ transplantation and from a mother to her fetus [77 - 80]. It is estimated that 7 to 8 million people mainly in Mexico, Central and South America that have Chagas disease, with almost 12,500 deaths in 2006. Most patients are poor and unaware of conducting infection. Migration has increased the areas with Chagas disease to involve many European countries and the United States. These areas have also seen an increase in the years up to 2014 [78, 79]. Chagas disease has two stages: acute and chronic. The acute stage has no or very mild symptoms like; fever and malaise. After the acute stage, the disease undergoes into remission and no other symptoms may appear for decades, till the disease progress into a chronic stage in approximately one-third of individuals. Ninety percent of the chronic cases develop heart disease and the other 10% show gastrointestinal complications [81]. The manifestations involving the heart include dilated cardiomyopathy, heart embolism, stroke, and congestive heart failure, which is the main cause of death in chronic Chagas disease [82, 83]. Chronic chagasic cardiomyopathy is characterized by a diffuse inflammatory reaction with mononuclear cell, leading to cardiac myocytolysis and mal-functioning [84]. Dysrhythmias are also common; with conduction abnormalities due to dense fibrosis of the bundle of Hiss. Autopsy may reveal marked bilateral ventricular enlargement, especially in the right side, with wall thinning, thrombi formation and apical aneurysms. Diffuse interstitial fibrosis, lymphocytes infiltrates and cardiomyocytes atrophy may be also present [85 - 87]. *T cruzi* parasites may be demonstrated by polymerase chain reaction (PCR) assays for parasites nucleic acid in the areas of focal inflammation, but the parasitic stages are rarely seen by microscopic examination [88].

There are limitations of drug therapy in both the acute and chronic stages. Heart transplantation in chagasic patients, with its high financial costs brings high potentiality of relapse of the acute *T. cruzi* infection due to the associated immunosuppressive therapy. On the other side, syngeneic bone marrow cell transplantation has been proved to induce a degree of repair and improvement of heart function in a number of studies on patients and animal models of ischemic cardiopathy [89]. This has shaded the light on the potentiality of using bone marrow cell transplantation in treating chronic chagasic cardiomyopathy. Transplantation of bone marrow mononuclear cells in mouse models of Chagas disease were found to be effective in reducing myocarditis and fibrosis within 6 months after transplantation. A marked repair with apoptosis of inflammatory cell infiltrate was observed, whether the bone marrow cells were transplanted from normal mice or even chagasic mice [90]. Also Goldenberg *et al.* reported a therapeutic and prophylactic effect of bone marrow mononuclear cells transplantation in *T. cruzi* induced ventricular dilatation of mice [91]. The mechanisms by which autologous BMC transplant regulates the immune-inflammatory processes in these cases are unknown, but lately, various reports witnessed a strong evidence of an autoimmune pathogenic element against myocardial antigens in chronic chagasic cardiomyopathy [89]. Thus, the benefits of BMC transplantation may originate from the regulation of pathological immune responses through inhibition of T-cell responses, in addition to tissue remodeling [92, 93]. However, although there was not an evidence of an increase in blood or heart parasitism in the BMC-transplanted animals to indicate immune down-regulation after transplantation, TNF-α producing cells were detected in both hearts of the mice and the patients with chagasic cardiomyopathy [94]. Hence, the reduced fibrosis after BMC transplantation can still be ascribed to its effect on decreasing the number of TNF- α producing cells in the heart [95].

Repeated injections of granulocyte colony-stimulating factor (G-CSF), to stimulate stem cell mobilization from the bone marrow, decreases the inflammation and fibrosis in the murine models of chagasic heart [96, 97]. A significant rise in T-regulatory cells, IL-10 and TGF-β and an obvious reduction in INF-γ and TNF-α were observed in the hearts of G-CSF-treated mice with simultaneous reduction of parasitism [97]. In 2013, Larocca *et al.* reported no improvement in arrhythmias and cardiovascular function in C57BL/6 mice with chronic chagasic cardiopathy after intraperitoneal treatment with mesenchymal human stem cells from adipose tissue (STAT) injection. However, heart sections revealed a significant reduction in inflammatory cells and fibrotic areas, in comparison with heart sections from chagasic animals treated with just Dulbecco/Vogt modified Eagle's minimal essential medium as control [98]. In Wistar rats, simultaneous autologous transplantation of cultured mesenchymal stem cells and skeletal myoblasts were found to significantly increase the ejection fraction volume and decrease the left ventricular end diastolic and systolic volume, indicating that co-transplant of stem cells and skeletal myoblasts is strongly effective in case of Chagas disease ventricular dysfunction [99]. Cardiac mesenchymal stem cells (CMSCs) have adipogenic, osteogenic, and chondrogenic differentiation potentials. Moreover, they could also express endothelial cell and cardiomyocyte characters in certain culture conditions and displayed immunosuppressive activity *in vitro*. To investigate the therapeutic potential of CMSCs in a murine experimental model of chronic Chagas disease, mice were intramyocardially injected with CMSCs, guided by echocardiograph in the left ventricle. CMSC-treated mice had a significantly decreased number of inflammatory cells, but without a reduction in the fibrotic area up to two months after treatment [99].

Based on previous studies, Soares and Ribeiro dos Santos decided to conduct a phase I clinical trial of autologous BMC therapy in 28 patients with Chagasic heart failure [100]. Bone marrow cells transplantation resulted in a significant improvement in these patients, suggesting that intracoronary injection of bone marrow mononuclear cells is possible and can be potentially safe and beneficial in these patients [101, 102]. This can be explained by the fact that chronic chagasic cardiomyopathy is specifically characterized by diffuse inflammation, in comparison to other causes of myocardial infarction and dilated cardiomyopathy, which shows a limited or no attraction to inflammatory cells to the site of lesion. Moreover, in chagasic cardiomyopathy, fibrosis spreads all over the heart, whereas in infarction fibrosis is limited to the insulted area, resulting in a difficulty in cells penetration into the damaged tissue [103].

Subsequent to the promising results of phase I clinical trial; a multicenter randomized trial was performed on a larger scale of chronic chagasic cardio-myopathy patients, aged between 18 and 75 years. Cases were randomized between receiving an intracoronary injection of autologous bone marrow-derived mononuclear cells (BMNCs) and only placebo. Unfortunately, no significant difference was recorded in the measured parameters between the study groups. Ribeiro Dos Santos *et al.* documented that intracoronary injection of autologous BMNCs could not improve chronic chagasic cardiomyopathy [104]. In order to reach an efficient treatment, certain points were needed to be put into consideration, including the optimal cell population, the number of cells and the doses required [105]. As many transplanted cells die just a few days after transplantation, measures required to prolong the survival and engraftment of these cells need to be investigated. Also, more researches on the molecular and cellular mechanisms, that drive the cells to injured sites, may help in developing new strategies for the treatment of Chagas heart disease. If secretion of soluble mediators and endogenous stem cells stimulation were proved to be the main effective mechanisms in cellular therapy of heart diseases, then cell therapy may be associated with or even replaced by the delivery of paracrine factors to repair the damaged chagasic heart [106].

However, the previous results collectively; encourage paying more efforts to determine the feasibility and safety of stem cell therapy in Chagas disease. Perhaps earlier intervention will be more efficient in treating or blocking the progress of the disease than late intervention, using the same protocol [107, 108].

4. PARASITIC STEM CELLS: CAN THEY BE THE TARGET OF ANTI-PARASITIC THERAPY?

A kind of adult pluripotent stem cells, called neoblasts, like those found in the human beings, , have been discovered in the long-lived free-living flatworms "planarians". Planaria are tiny flatworms that live in fresh and marine water, attached to plants all over the world. If one of these creatures was cut into pieces, each piece can grow into a whole organism alone. Hence, a single neoblast can give rise to all cell types and organs in the body of planarian. Neoblasts are abundant in the mesenchyme, where they continuously divide to drive long- term homeostatic tissue, and maintain the entire animal again within weeks [109, 110]. Presence of similar stem cells in human parasites can direct scientists to a new approach in the continuous battle against debilitating human parasites, like schistosomiasis and the neoplasm-like parasitic disease "alveolar echino-cocchosis".

4.A. Schistosomiasis

Nothing has been known about similar cells in the notoriously long-lived parasites of schistosomiasis. In 2013, Collin *et al*. described a group of neoblast-like cells in the trematode "*Schistosoma mansoni*" [111]. Transmission of these parasites relies on a stem cell-driven, clonal expansion inside their snail intermediate host [112]. This discovery can be of great importance as such cells can not only be new drug targets against schistosomiasis in the human host, or a control measure against the molluscum host in the external environment, but they can be a potentially vaccine target as well. Schistosomiasis is one of the most chronic and prevalent human parasitic diseases. *Schistosoma mansoni* infect mainly the vascular plexus around the intestine and may also infect the one around the urinary tract. Symptoms appear as abdominal pain, dysentry or terminal haematurea. In chronic patients, the case progress to hepatic fibrosis, ascitis, renal failure, infertility and may be squamous cell carcinoma of the urinary bladder. In children it causes stunted growth [113]. About 230 million people worldwide are infected with this disease with an annual death rate; 250,000 per year [114]. The schistosomes life cycle alternates between asexual and sexual generations in invertebrate intermediate and vertebrate definitive hosts, respectively. The life cycle initiates when mature eggs are excreted from a mammal into freshwater. Eggs hatch, releasing miracidia that seek out and penetrate the hosting snail. Entry into the snail triggers a series of morphological, physiological, and chemical changes, followed by a clonal expansion in the form of sporocysts inside the snail, which ultimately produces many infective cercariae. Cercariae exit the snail to the fresh water, where they become attracted by vibration to the mammal host and burrow through the skin, leaving their tails in the epidermis. In the mammal, the new stage "schistosomula" keeps migrating in the cardiovascular system, till reach the liver, where development to adulthood takes place, and sexual reproduction starts. Thus, asexual amplification inside of the snail is vital for the propagation of schistosomes [115]. Though adult schistosomes do not replicate within the host, they can survive as long as 30 years in the human body. The pluripotent stem cells in *Schistosoma mansoni* described by Collins *et al.* in Nature, can proliferate and differentiate into many germ layers derivatives. This explains why adult worms can change their growth pattern according to the immune response and the male-female worms pairing [111]. They can also regenerate damaged tissues following sub-lethal doses of praziquantel. Schistosome neoblast-like cells express a fibroblast growth factor receptor ortholog may help to reveal how the developmental plasticity of schistosomes for all those years or even decades inside their host, and suggest that these parasites can utilize different developmental programs to react against the external stimuli [111, 116]. According to Newmark *et al.* [117]; to track stem cells in *Schistosoma*, worms were grown in culture and fluorescent tags were added to label the newly replicated DNA. After that, they saw a separate punch of proliferating cells in the worm's body. Those cells were isolated and studied separately. They were typical stem cells with their large nucleus and little cytoplasm that left limited room for functional organelles. The cells were often found to divide giving rise to two sets of cells: one continued to divide and another did not. This means that they make more stem cells; then, many stem cells would have asymmetric division. After division, labeled cells were seen in the schistosomes intestines and muscles. Hence, those schistosomes cells were behaving as stem cells and even differentiate into other cell types [112, 118]. Wang *et al.* hypothesize that the germinal cells found during the asexual amplification in the snail may have a similar molecular program like the adult stage [112].

A population of totipotent stem cells, historically called 'germinal cells', is thought to underlie this unique intramolluscan amplification by undergoing multiple rounds of proliferation and de novo embryogenesis in the absence of fertilization. The proliferating cells in sporocysts express many conserved stem cell genes. Comparison between sporocyst germinal cells described in this study and somatic stem cells recently identified in adult schistosomes revealed a significant molecular similarity between these germinal cells in schistosome larvae and somatic stem cells in schistosome adults. This suggests that these stem cells may persist throughout the entire schistosome life cycle and links embryonic development and homeostatic tissue maintenance in these parasites [116]. The neoblast-like stem cells in the molluscum stage "sporocyst" possess a molecular signature similar to that of neoblasts in the free-living flatworms "planarians". This suggests that *Schistosoma* developmental programs inherited from their free-living ancestors may have enabled the evolution of the complex life cycles in trematode. These pluripotent cells may help the worms to successfully adapt to obligate parasitism, by the rapid expansion of the infective population and the long-term tissue maintenance in the hostile environment inside the intermediate host and definitive host [119, 120]. Future studies about the role of the neoblast-like cells will have important implications to understand their biological roles in schistosomes and to target them with a suitable new anti- helminthic therapy. If a drug or vaccine succeeds to eliminate the schistosome stem cells in the sporocyst and adult stages, worms would either die off or at least become sterile according to the targeted stem cells. Drug or vaccine that induces worm infertility would achieve a stunting therapeutic outcome. It would hinder the parasite transmission and stop the egg-induced tissue pathology, which is the main precipitating factor of tissue damage and complications in schistosomiasis [121].

4.B. Alveolar Ecchinochoccosis

Alveolar ecchinochoccosis is a fatal disease caused by the larvae of *Echinococcus multilocularis*, which is a cestodal tapeworm. Infection of the humans starts by ingestion of the parasite eggs that contain the oncosphere larva. Once eggs hatch in the small intestine, this oncosphere penetrates the intestinal epithelium and reach different host organs, mainly the liver, to grow into a larva. The larvae are characterized by being a continuously spreading infiltrative mass that keeps spreading into the organs, to metastasize locally and through lymphatics. This unlimited growth is driven by the stem cells in the germinal layer.

For effective cure of this disease the parasite stem cells have to be eliminated. Unfortunately, the current treatment is benzimidazoles, which is parasitostatic only. Benzimidazoles don't kill the germinative cell, hence once the treatment is discontinued; recurrence occurs [122 - 124]. After isolation and cultivation of the germinal layer stem cells, they showed the typical characters of stem cells with production of new metacestode tissue with several gene expressions of typical components of the germ line multipotency program of metazoan stem cells [125, 126]. Based on that, Schuber *et al*. have succeeded in describing a parasite enzyme; EmPlk1 [127]. This novel enzyme is responsible for the regulation of germinative cell proliferation. Low concentrations of BI 2536, which is a compound originally designed to inhibit the human ortholog of EmPlk1, have been found to be able to inhibit the parasite enzyme EmPlk1, and subsequently, eliminate the germinative cell in the larval vesicles *in vitro*, yielding the parasite no longer able to proliferate. Thus, BI 2536 and other compounds targeting EmPlk1 enzyme in the parasite stem cells, can be very promising drugs if combined with albendazole in the treatment of this lethal cancer-like disease, caused by *Echinococcus multilocularis* [127].

5. PARASITES: CAN THEY BE TRIGGERS FOR STEM CELL FACTOR PRODUCTION AND STEM CELL CULTURE?

There are certain immunological changes that follow parasitic infections, especially intestinal helminths. These changes occur mainly through the stimulation of Th2 immune response. They include; increased serum IgE, eosinophilia and mast cell hyperplasia. Mastocytosis occurs through the stimulation of stem cell factor (SCF) production. SCF is considered an important cytokine in the process of haematopoiesis due to the abundant expression of its receptor (C-kit) on hematopoietic progenitor cells like; erythroblasts, myeloblasts, and megakaryocytes. Mast cell is the only mature and terminally differentiated cell that expresses the c-Kit receptor and subsequently mastocytosis occurs after an intestinal infection by a helminth. SCF may be used with other cytokines to enhance human stem cells (HSCs) and hematopoietic progenitor cultures. The expansion of these cells *in-vitro* allows progress in bone marrow transplantation, where HSCs are transferred to replenish blood formation in the recipient. Thus, studies on helminths antigens (like *Trichinella spiralis, Necator braziliensis* and *Schistosoma mansoni*) and their impact on SCF, can consist a potential trigger factor to enhance SCF production and subsequently be beneficial in the process of stem cell culture and bone marrow transplantation [128].

Mast cells are first located in the bone marrow as cell precursors. After that, they leave the bone marrow as progenitors, and then get mature in the periphery e.g. small intestine and peripheral circulation. Throughout mastopoiesis, mast cells loose cytokine binding sites to become mature cells with undetectable expression of IL-3R, GM-CSFR or IL-8R. In contrast to other myeloid cells, mast cells keep expressing of stem cell factor (SCF) receptor/c- kit during both mastopoiesis and maturity. Furthermore, SCF (the ligand of c-kit), plays a role in both induction of human mast cells differentiation from their progenitor cells and then up-regulation of their effector functions after maturity [158]. In 1990, the cloning and characterization of SCF was described. Stem Cell Factor (also known as C-kit ligand "KL" or steel factor) binds to the C-Kit receptor presented on certain cells. It's secreted by bone marrow stromal cells endothelium and fibroblasts, as a membrane-bound form which on cleavage, it releases the soluble form. Both forms promote colony formation of bone marrow cells. However, membrane-bound SCF is more effective in hematopoieses *in vivo*, which indicates the necessity of cellular interactions between hematopoietic and stromal cells. It's expressed in human blood at a serum level of 3.3 ng/ml [129]. Stem Cell Factor plays an important role in hematopoiesis, spermatogenesis, melanogenesis, and possibly CNS development. C-Kit receptor (CD 117) is expressed in HSCs, mast cells, melanocytes, germ cells and also expressed in hematopoietic progenitor cells like; erythroblasts, myeloblasts, and megakaryocytes. During development; SCF participates in melanocytes and hematopoietic stem cells (HSCs) localization. The fetal liver and bone marrow are the main sites for hematopoiesis. They express SCF to attract HSCs to their stem cell niche, and then contribute in self- renewal and maintenance of HSCs. Hence, mutations in C-Kit receptor end up with anemia and a decrease in the fertility and pigmentation [130]. It is believed that concentration gradients of SCF can allow HSCs to return to their niches, after they regularly leave the bone marrow to the general circulation during adulthood. Fetal HSCs were found to be six times more sensitive to SCF than HSCs from adults in the maximum survival concentrations. SCF also increases the survival of various hematopoietic progenitor cells, such as megakaryocyte progenitors, and act as chemotactic factor to hematopoietic progenitor cells *in vitro* [131].

Exception for mast cells, C-Kit expression keeps decreasing, till it completely disappears on hematopoietic cells maturation [132]. Hence, mice with SCF or C-Kit mutations do have serious defects in mast cells production with less than 1% of the normal levels. Conversely, the injection of SCF increases mast cell recruitment, proliferation and adhesion in the site of injection over a hundred times [133].

SCF may be used along with other cytokines in HSCs and hematopoietic progenitor's cultures. *In vitro* expansion of these cells will allow more advances in bone marrow transplantation, for re-establishment of blood formation. Unfortunately, one of the problems of injecting SCF for therapeutic purposes is mast cells activation, which results in allergic-like symptoms due to histamine release. Meanwhile, IL-3 and IL-9 have been identified as regulators of mast cell growth and differentiation, in addition to SCF [134]. SCF plays a critical role in parasites induced mast cell hyperplasia. Intestinal mastocytosis is a strong feature in most of intestinal nematode infections, like *Nippostrongylus brasiliensis* and *Trichinella spiralis*, where SCF triggers the unipotent stem cells "Mast cell precursor" to enhance mast cell growth in the mucosa and connective tissue [135, 136]. A variety of immunological changes characteristically accompany parasitic infections, include; eosinophilia and increased serum IgE in addition to mast cell hyperplasia. Normally, activation of mast cells occurs *via* immunoglobulin E (IgE) stimulation. Antigen binding by high-affinity IgE receptor FcεRI induces a stream of signals and ultimately the release of mast cells mediators. This mechanism has been initially described in the context of allergy, and parasitic infections as well [137]. Mast cells play an important role in the immune response against both intestinal and tissue parasites. For example, they accelerate worm expulsion from the gut as in Anclystoma duodenale, *Trichinella spiralis* and *Strongyloides stercoralis*, and limit the tissue burden of parasites. In mast cell–deficient mice larger *Leishmania major* lesions develop with increased parasite load and less inflammatory manifestations. Moreover, in human, cases of atypical leishmaniasis that mimic cutaneous masocytosis have been reported. Finally regarding malaria, a critical role for mast cell–derived TNF has been demonstrated in limiting parasitaemia after reconstitution of mast cells in murine models deficient in mast cells [138, 139]. Parasitic induced mastocytosis is under the control of Th2 cells with SCF as a cofactor. However, in absence of Th2 cytokines, SCF alone can't stimulate the clonal growth of primitive mast cell progenitors. SCF has potent growth-promoting effects when combined with IL-3, IL-4, and IL-10 [140]. Optimal mast cell colony formation occurs when both IL-4 and IL-10 are combined with SCF. That was proved by experimental studies on *Trichinella spiralis* intestinal infections in rodents, in which an 85% reduction of mastocytosis in gut mast cells has occurred after *in vivo* mAb neutralization of these cytokines [141]. However, IL-4 was found to be more efficient than IL-10 in co-stimulating SCF-dependent mast cell progenitors colony formation and sustaining their proliferation. The mast cells generated in colonies stimulated by IL-4, IL-10, and SCF were found to produce high levels of histamine (6-8 pg per cell) [142]. Researches on *S. mansoni* infected rats have similarly showed depletion of mast cells and a rapid decrease in serum level of rat mucosal mast cell protease II (RMCP II) after treatment with sheep anti-SCF antibody. Correspondingly, the number of mucosal mast cells and RMCP II levels in the small intestine and liver were significantly reduced. However, eosinophil recruitment to the liver and intestine in S. mansoni infection was not SCF dependent [136, 141, 143]. In 1995, Newlands *et al.* had also documented a significant role of SCF in mast cell hyperplasia after *N. brasiliensis* infections in rats [144]. However, treatment of infected rats with a polyclonal sheep anti-rat

SCF did not prevent or delay expulsion of the parasite from the intestine. Indeed, it had affected the parasite fecundity. These differences are probably related to the parasite annd suggests that mast cells are not important in resistance against the hook worms.

Although the main source of SCF production is the bone marrow, recent work has demonstrated SCF mRNA in the intestine of normal mice [145]. However, the site of SCF production during parasitic infection is still unknown till present. According to Donaldson *et al*. during *Trichinella spiralis* infection both the bone marrow and the gut remain the two possible sources for SCF production. Though SCF can be secreted in the local microenvironment or act through its membrane form, circulating SCF couldn't be detected in Donaldson *et al*. study. This suggests that; locally mediated effect has the most probable priming effect during infection with intestinal nematode [141]. Collectively, these results ascertain the critical role of SCF-C-kit system in the protective response against parasitic infection by mastocytosis [146, 147] and emphasize the stimulatory effect of parasites on SCF production, and the subsequent effect on HSCs and hematopoietic progenitors.

CONFLICT OF INTEREST

The author confirms that author has no conflict of interest to declare for this publication.

ACKNOWLEDGEMENTS

Declared none.

REFERENCES

[1]   Cox FE. History of human parasitology. Clin Microbiol Rev 2002; 15(4): 595-612.
        [http://dx.doi.org/10.1128/CMR.15.4.595-612.2002] [PMID: 12364371]

[2]   Crum-Cianflone NF. Bacterial, fungal, parasitic, and viral myositis. Clin Microbiol Rev 2008; 21(3): 473-94.
        [http://dx.doi.org/10.1128/CMR.00001-08] [PMID: 18625683]

[3]   WHO. Malaria Fact sheet NA°94 WHO March 2014 , [Retrieved 28 August 2014];

[4]   Khuroo MS. Ascariasis. Gastroenterol Clin North Am 1996; 25(3): 553-77.
        [http://dx.doi.org/10.1016/S0889-8553(05)70263-6] [PMID: 8863040]

[5]   Hassl A. [An introduction into bioethics and parasitology]. Wien Klin Wochenschr 2006; 118(19-20) (Suppl. 3): 33-6.
        [http://dx.doi.org/10.1007/s00508-006-0682-2] [PMID: 17131238]

[6]   Alemu A, Shiferaw Y, Getnet G, Yalew A, Addis Z. Opportunistic and other intestinal parasites among HIV/AIDS patients attending Gambi higher clinic in Bahir Dar city, North West Ethiopia. Asian Pac J Trop Med 2011; 4(8): 661-5.
        [http://dx.doi.org/10.1016/S1995-7645(11)60168-5] [PMID: 21914548]

[7]   Day T. Parasite transmission modes and the evolution of virulence. Evolution 2001; 55(12): 2389-400.
        [http://dx.doi.org/10.1111/j.0014-3820.2001.tb00754.x] [PMID: 11831655]

[8]   Singh G, Sehgal R. Transfusion-transmitted parasitic infections. Asian J Transfus Sci 2010; 4(2): 73-7.
        [http://dx.doi.org/10.4103/0973-6247.67018] [PMID: 20859503]

[9]   Brockhurst MA, Buckling A, Poullain V, Hochberg ME. The impact of migration from parasite-free patches on antagonistic host-parasite coevolution. Evolution 2007; 61(5): 1238-43.
        [http://dx.doi.org/10.1111/j.1558-5646.2007.00087.x] [PMID: 17492974]

[10]   MacDonald AS, Araujo MI, Pearce EJ. Immunology of parasitic helminth infections. Infect Immun 2002; 70(2): 427-33.
        [http://dx.doi.org/10.1128/IAI.70.2.427-433.2002] [PMID: 11796567]

[11]   Hunter PR, Nichols G. Epidemiology and Clinical Features of Cryptosporidium Infection in Immunocompromised Patients 2002.
        [http://dx.doi.org/10.1128/CMR.15.1.145-154.2002]

[12]   Shirley DA, Moonah SN, Kotloff KL. Burden of disease from cryptosporidiosis. Curr Opin Infect Dis 2012; 25(5): 555-63.
        [http://dx.doi.org/10.1097/QCO.0b013e328357e569] [PMID: 22907279]

[13]   MacKenzie WR, *et al.* A massive outbreak in Milwaukee of *Cryptosporidium* infection transmitted through the public water supply. N Engl J Med 1994; 331: 67-161.

[14]   Xiao L, Morgan UM, Fayer R, Thompson RC, Lal AA. *Cryptosporidium* systematics and implications for public health. Parasitol Today (Regul Ed) 2000; 16(7): 287-92.
        [http://dx.doi.org/10.1016/S0169-4758(00)01699-9] [PMID: 10858647]

[15]   Nachbaur D, Kropshofer G, Feichtinger H, Allerberger F, Niederwieser D. Cryptosporidiosis after CD34-selected autologous peripheral blood stem cell transplantation (PBSCT). Treatment with paromomycin, azithromycin and recombinant human interleukin-2. Bone Marrow Transplant 1997; 19(12): 1261-3.
        [http://dx.doi.org/10.1038/sj.bmt.1700826] [PMID: 9208124]

[16]   Gentile G, Venditti M, Micozzi A, *et al.* Cryptosporidiosis in patients with hematologic malignancies. Rev Infect Dis 1991; 13(5): 842-6.
        [http://dx.doi.org/10.1093/clinids/13.5.842] [PMID: 1962096]

[17]   Manivel C, Filipovich A, Snover DC. Cryptosporidiosis as a cause of diarrhea following bone marrow transplantation. Dis Colon Rectum 1985; 28(10): 741-2.
        [http://dx.doi.org/10.1007/BF02560294] [PMID: 3902409]

[18]   Martino P, Gentile G, Caprioli A, *et al.* Hospital-acquired cryptosporidiosis in a bone marrow transplantation unit. J Infect Dis 1988; 158(3): 647-8.
[http://dx.doi.org/10.1093/infdis/158.3.647] [PMID: 3045217]

[19]   Nachbaur D, Kropshofer G, Feichtinger H, Allerberger F, Niederwieser D. Cryptosporidiosis after CD34-selected autologous peripheral blood stem cell transplantation (PBSCT). Treatment with paromomycin, azithromycin and recombinant human interleukin-2. Bone Marrow Transplant 1997; 19(12): 1261-3.
[http://dx.doi.org/10.1038/sj.bmt.1700826] [PMID: 9208124]

[20]   Martino P, Gentile G, Caprioli A, *et al.* Hospital-acquired cryptosporidiosis in a bone marrow transplantation unit. J Infect Dis 1988; 158(3): 647-8.
[http://dx.doi.org/10.1093/infdis/158.3.647] [PMID: 3045217]

[21]   Beaman MH, *et al. Toxoplasma gondii.* In: Mandell GL, Bennett JE, Dolin R, Eds. Principles and practice of infectious diseases. New York: Churchill Livingston 1995; pp. 2455-75.

[22]   Chandrasekar PH, Momin F. Disseminated toxoplasmosis in marrow recipients: a report of three cases and a review of the literature. Bone Marrow Transplant 1997; 19(7): 685-9.
[http://dx.doi.org/10.1038/sj.bmt.1700736] [PMID: 9156245]

[23]   Martino R, Maertens J, Bretagne S, *et al.* Toxoplasmosis after hematopoietic stem cell transplantation. Clin Infect Dis 2000; 31(5): 1188-95.
[http://dx.doi.org/10.1086/317471] [PMID: 11073751]

[24]   Derouin F, Devergie A, Auber P, *et al.* Toxoplasmosis in bone marrow-transplant recipients: report of seven cases and review. Clin Infect Dis 1992; 15(2): 267-70.
[http://dx.doi.org/10.1093/clinids/15.2.267] [PMID: 1520761]

[25]   Sing A, Leitritz L, Roggenkamp A, *et al.* Pulmonary toxoplasmosis in bone marrow transplant recipients: report of two cases and review. Clin Infect Dis 1999; 29(2): 429-33.
[http://dx.doi.org/10.1086/520228] [PMID: 10476754]

[26]   Costa JM, Pautas C, Ernault P, Foulet F, Cordonnier C, Bretagne S. Real-time PCR for diagnosis and follow-up of *Toxoplasma* reactivation after allogeneic stem cell transplantation using fluorescence resonance energy transfer hybridization probes. J Clin Microbiol 2000; 38(8): 2929-32.
[PMID: 10921953]

[27]   Sirvent-von Bueltzingsloewen A, Marty P, Rosenthal E, *et al.* Visceral leishmaniasis: a new opportunistic infection in hematopoietic stem-cell-transplanted patients. Bone Marrow Transplant 2004; 33(6): 667-8.
[http://dx.doi.org/10.1038/sj.bmt.1704396] [PMID: 14730334]

[28]   Dujardin JC, Campino L, CaA avate C, *et al.* Spread of vector-borne diseases and neglect of Leishmaniasis, Europe. Emerg Infect Dis 2008; 14(7): 1013-8.
[http://dx.doi.org/10.3201/eid1407.071589] [PMID: 18598618]

[29]   Handman E, Bullen DV. Interaction of *Leishmania* with the host macrophage. Trends Parasitol 2002; 18(8): 332-4.
[http://dx.doi.org/10.1016/S1471-4922(02)02352-8] [PMID: 12377273]

[30]   Rebelatto CK, Aguiar AM, MoretA o MP, *et al.* Dissimilar differentiation of mesenchymal stem cells

from bone marrow, umbilical cord blood, and adipose tissue. Exp Biol Med (Maywood) 2008; 233(7): 901-13.
[http://dx.doi.org/10.3181/0712-RM-356] [PMID: 18445775]

[31]   Choi YS, Cha SM, Lee YY, Kwon SW, Park CJ, Kim M. Adipogenic differentiation of adipose tissue derived adult stem cells in nude mouse. Biochem Biophys Res Commun 2006; 345(2): 631-7.
[http://dx.doi.org/10.1016/j.bbrc.2006.04.128] [PMID: 16696950]

[32]   Allahverdiyev AM, Bagirova M, Elcicek S, *et al.* Adipose tissue-derived mesenchymal stem cells as a new host cell in latent leishmaniasis. Am J Trop Med Hyg 2011; 85(3): 535-9.
[http://dx.doi.org/10.4269/ajtmh.2011.11-0037] [PMID: 21896818]

[33]   Mazo M, Abizanda G, Pelacho B, *et al.* Transplantation of adipose derived stromal cells is associated with functional improvement in a rat model of chronic myocardial infarction. Eur J Heart Fail 2008; 10(5): 454-62.
[http://dx.doi.org/10.1016/j.ejheart.2008.03.017] [PMID: 18436478]

[34]   Scharton-Kersten T, Afonso LC, Wysocka M, Trinchieri G, Scott P. IL-12 is required for natural killer cell activation and subsequent T helper 1 cell development in experimental leishmaniasis. J Immunol 1995; 154(10): 5320-30.
[PMID: 7730635]

[35]   Antinori S, Cascio A, Parravicini C, Bianchi R, Corbellino M. Leishmaniasis among organ transplant recipients. Lancet Infect Dis 2008; 8(3): 191-9.
[http://dx.doi.org/10.1016/S1473-3099(08)70043-4] [PMID: 18291340]

[36]   Cell Press. How The Malaria Parasite Hijacks Human Red Blood Cells. Science Daily from www.sciencedaily.com/releases/2008/07/080708155615.htm , 2008 July 10; [Retrieved September 24, 2014];

[37]   Leiby DA. Making sense of malaria. Transfusion 2007; 47(9): 1573-7.
[http://dx.doi.org/10.1111/j.1537-2995.2007.01418.x] [PMID: 17725719]

[38]   Chiche L, Lesage A, Duhamel C, *et al.* Posttransplant malaria: first case of transmission of *Plasmodium falciparum* from a white multiorgan donor to four recipients. Transplantation 2003; 75(1): 166-8.
[http://dx.doi.org/10.1097/00007890-200301150-00031] [PMID: 12544892]

[39]   Owusu-Ofori AK, Parry C, Bates I. Transfusion-transmitted malaria in countries where malaria is endemic: a review of the literature from sub-Saharan Africa. Clin Infect Dis 2010; 51(10): 1192-8.
[http://dx.doi.org/10.1086/656806] [PMID: 20929356]

[40]   Wickramasinghe SN, Abdalla SH. Blood and bone marrow changes in malaria. Best Pract Res Clin Haematol 2000; 13(2): 277-99.
[http://dx.doi.org/10.1053/beha.1999.0072] [PMID: 10942626]

[41]   Stramer SL, Hollinger FB, Katz LM, *et al.* Emerging infectious disease agents and their potential threat to transfusion safety. Transfusion 2009; 49 (Suppl. 2): 1S-29S.
[http://dx.doi.org/10.1111/j.1537-2995.2009.02279.x] [PMID: 19686562]

[42]   Mejia R, Booth GS, Fedorko DP, *et al.* Peripheral blood stem cell transplant-related *Plasmodium falciparum* infection in a patient with sickle cell disease. Transfusion 2012; 52(12): 2677-82.

[http://dx.doi.org/10.1111/j.1537-2995.2012.03673.x] [PMID: 22536941]

[43]   Dharmasena F, Gordon-Smith EC. Transmission of malaria by bone marrow transplantation. Transplantation 1986; 42(2): 228.
[http://dx.doi.org/10.1097/00007890-198608000-00027] [PMID: 3526663]

[44]   Tran V-B, Tran VB, Lin KH. Malaria infection after allogeneic bone marrow transplantation in a child with thalassemia. Bone Marrow Transplant 1997; 19(12): 1259-60.
[http://dx.doi.org/10.1038/sj.bmt.1700822] [PMID: 9208123]

[45]   LefrA"re F, Besson C, Datry A, *et al.* Transmission of *Plasmodium falciparum* by allogeneic bone marrow transplantation. Bone Marrow Transplant 1996; 18(2): 473-4.
[PMID: 8864468]

[46]   Raina V, Sharma A, Gujral S, Kumar R. Plasmodium vivax causing pancytopenia after allogeneic blood stem cell transplantation in CML. Bone Marrow Transplant 1998; 22(2): 205-6.
[http://dx.doi.org/10.1038/sj.bmt.1701299] [PMID: 9707032]

[47]   Hsieh MM, Kang EM, Fitzhugh CD, *et al.* Allogeneic hematopoietic stem-cell transplantation for sickle cell disease. N Engl J Med 2009; 361(24): 2309-17.
[http://dx.doi.org/10.1056/NEJMoa0904971] [PMID: 20007560]

[48]   Fasano RM, Monaco A, Meier ER, *et al.* RH genotyping in a sickle cell disease patient contributing to hematopoietic stem cell transplantation donor selection and management. Blood 2010; 116(15): 2836-8.
[http://dx.doi.org/10.1182/blood-2010-04-279372] [PMID: 20644109]

[49]   Wickramasinghe SN, Abdalla SH. Blood and bone marrow changes in malaria. Best Pract Res Clin Haematol 2000; 13(2): 277-99.
[http://dx.doi.org/10.1053/beha.1999.0072] [PMID: 10942626]

[50]   Desakorn V, Dondorp AM, Silamut K, *et al.* Stage-dependent production and release of histidine-rich protein 2 by *Plasmodium falciparum*. Trans R Soc Trop Med Hyg 2005; 99(7): 517-24.
[http://dx.doi.org/10.1016/j.trstmh.2004.11.014] [PMID: 15876442]

[51]   Zhang Y, Mi JY, Rui YJ, Xu YL, Wang W. Stem cell therapy for the treatment of parasitic infections: is it far away? Parasitol Res 2014; 113(2): 607-12.
[http://dx.doi.org/10.1007/s00436-013-3689-4] [PMID: 24276645]

[52]   Xu HJ, Qian H, Zhu W, *et al.* [Inhibition of culture supernatant of mesenchymal stem cells on macrophages RAW264.7 activated by soluble egg antigen of Schistosoma japonicum]. Zhongguo Ji Sheng Chong Xue Yu Ji Sheng Chong Bing Za Zhi 2011; 29(6): 425-30.
[PMID: 24822341]

[53]   Wilson MS, Mentink-Kane MM, Pesce JT, Ramalingam TR, Thompson R, Wynn TA. Immunopathology of schistosomiasis. Immunol Cell Biol 2007; 85(2): 148-54.
[http://dx.doi.org/10.1038/sj.icb.7100014] [PMID: 17160074]

[54]   Wu W, Huang Y. Application of praziquantel in schistosomiasis japonica control strategies in China. Parasitol Res 2013; 112(3): 909-15.
[http://dx.doi.org/10.1007/s00436-013-3303-9] [PMID: 23358736]

[55]   El-Ansary M, Abdel-Aziz I, Mogawer S, *et al.* Phase II trial: undifferentiated *versus* differentiated

autologous mesenchymal stem cells transplantation in Egyptian patients with HCV induced liver cirrhosis. Stem Cell Rev 2012; 8(3): 972-81.
[http://dx.doi.org/10.1007/s12015-011-9322-y] [PMID: 21989829]

[56]   Yu Y, Lu L, Qian X, *et al.* Antifibrotic effect of hepatocyte growth factor-expressing mesenchymal stem cells in small-for-size liver transplant rats. Stem Cells Dev 2010; 19(6): 903-14.
[http://dx.doi.org/10.1089/scd.2009.0254] [PMID: 20025519]

[57]   Xu H, Qian H, Zhu W, *et al.* Mesenchymal stem cells relieve fibrosis of Schistosoma japonicum-induced mouse liver injury. Exp Biol Med (Maywood) 2012; 237(5): 585-92.
[http://dx.doi.org/10.1258/ebm.2012.011362] [PMID: 22678013]

[58]   Chawla YK. Trends and research in viral hepatitis. Indian J Med Res 2008; 127: 200.

[59]   Abo Raya DM, *et al.* Effct of stem cell transplantation on hepatic fibrosis in schistosoma mansoni-infected mice. [PhD thesis] 2012.

[60]   Richter J, Franken G, Mehlhorn H, Labisch A. What is the evidence for the existence of Plasmodium ovale hypnozoites? Parasitol Res 2010; 107(6): 1285-90.
[http://dx.doi.org/10.1007/s00436-010-2071-z] [PMID: 20922429]

[61]   Ghosh K, Ghosh K. Pathogenesis of anemia in malaria: a concise review. Parasitol Res 2007; 101(6): 1463-9.
[http://dx.doi.org/10.1007/s00436-007-0742-1] [PMID: 17874326]

[62]   Saei AA, Ahmadian S. Stem cell engineering might be protective against severe malaria. Biosci Hypotheses 2009; 2: 48-9.
[http://dx.doi.org/10.1016/j.bihy.2008.10.008]

[63]   Asami M, Owhashi M, Abe T, Nawa Y. Susceptibility of multipotent haemopoietic stem cell deficient W/Wv mice to Plasmodium berghei-infection. Immunol Cell Biol 1991; 69(Pt 5): 355-60.
[http://dx.doi.org/10.1038/icb.1991.51] [PMID: 1787005]

[64]   Belyaev NN, Brown DE, Diaz AI, *et al.* Induction of an IL7-R(+)c-Kit(hi) myelolymphoid progenitor critically dependent on IFN-gamma signaling during acute malaria. Nat Immunol 2010; 11(6): 477-85.
[http://dx.doi.org/10.1038/ni.1869] [PMID: 20431620]

[65]   Peterson VM, Madonna GS, Vogel SN. Differential myelopoietic responsiveness of BALB/c (Itys) and C.D2 (Ityr) mice to lipopolysaccharide administration and Salmonella typhimurium infection. Infect Immun 1992; 60(4): 1375-84.
[PMID: 1548063]

[66]   Weiss L, Johnson J, Weidanz W. Mechanisms of splenic control of murine malaria: tissue culture studies of the erythropoietic interplay of spleen, bone marrow, and blood in lethal (strain 17XL) Plasmodium yoelii malaria in BALB/c mice. Am J Trop Med Hyg 1989; 41(2): 135-43.
[PMID: 2774062]

[67]   Clark CR, Chen BD, Boros DL. Macrophage progenitor cell and colony-stimulating factor production during granulomatous schistosomiasis mansoni in mice. Infect Immun 1988; 56(10): 2680-5.
[PMID: 3138179]

[68]   Cotterell SE, Engwerda CR, Kaye PM. *Leishmania* donovani infection of bone marrow stromal macrophages selectively enhances myelopoiesis, by a mechanism involving GM-CSF and TNF-alpha.

Blood 2000; 95(5): 1642-51.
[PMID: 10688819]

[69] Mirkovich AM, Galelli A, Allison AC, Modabber FZ. Increased myelopoiesis during *Leishmania* major infection in mice: generation of ?~safe targets ?(tm), a possible way to evade the effector immune mechanism. Clin Exp Immunol 1986; 64(1): 1-7.
[PMID: 3488146]

[70] Engwerda CR, Good MF. A novel pathway of haematopoiesis revealed after experimental malaria infection. Immunol Cell Biol 2010; 88(7): 692-4.
[http://dx.doi.org/10.1038/icb.2010.89] [PMID: 20644560]

[71] Thakur RS, Tousif S, Awasthi V, *et al.* Mesenchymal stem cells play an important role in host protective immune responses against malaria by modulating regulatory T cells. Eur J Immunol 2013; 43(8): 2070-7.
[http://dx.doi.org/10.1002/eji.201242882] [PMID: 23670483]

[72] Kawada H, Fujita J, Kinjo K, *et al.* Nonhematopoietic mesenchymal stem cells can be mobilized and differentiate into cardiomyocytes after myocardial infarction. Blood 2004; 104(12): 3581-7.
[http://dx.doi.org/10.1182/blood-2004-04-1488] [PMID: 15297308]

[73] Sung JH, Yang HM, Park JB, *et al.* Isolation and characterization of mouse mesenchymal stem cells. Transplant Proc 2008; 40(8): 2649-54.
[http://dx.doi.org/10.1016/j.transproceed.2008.08.009] [PMID: 18929828]

[74] Pasare C, Medzhitov R. Toll pathway-dependent blockade of CD4+CD25+ T cell-mediated suppression by dendritic cells. Science 2003; 299(5609): 1033-6.
[http://dx.doi.org/10.1126/science.1078231] [PMID: 12532024]

[75] Bettelli E, Carrier Y, Gao W, *et al.* Reciprocal developmental pathways for the generation of pathogenic effector TH17 and regulatory T cells. Nature 2006; 441(7090): 235-8.
[http://dx.doi.org/10.1038/nature04753] [PMID: 16648838]

[76] Uccelli A, Moretta L, Pistoia V. Mesenchymal stem cells in health and disease. Nat Rev Immunol 2008; 8(9): 726-36.
[http://dx.doi.org/10.1038/nri2395] [PMID: 19172693]

[77] Rassi A Jr, Rassi A, Marin-Neto JA. Chagas disease. Lancet 2010; 375(9723): 1388-402.
[http://dx.doi.org/10.1016/S0140-6736(10)60061-X] [PMID: 20399979]

[78] Bonney KM. Chagas disease in the 21st century: a public health success or an emerging threat? Parasite 2014; 21: 11.
[http://dx.doi.org/10.1051/parasite/2014012] [PMID: 24626257]

[79] Chagas disease (American trypanosomiasis) Fact sheet N°340. World Health Organization March 2014 , 2014 [Retrieved 29 September 2014];

[80] BarcA n L, Luna C, Clara L, *et al.* Transmission of T. cruzi infection *via* liver transplantation to a nonreactive recipient for Chagas ?(tm) disease. Liver Transpl 2005; 11(9): 1112-6.
[http://dx.doi.org/10.1002/lt.20522] [PMID: 16123968]

[81] Zhang L, Tarleton RL. Parasite persistence correlates with disease severity and localization in chronic Chagas ?(tm) disease. J Infect Dis 1999; 180(2): 480-6.

[http://dx.doi.org/10.1086/314889] [PMID: 10395865]

[82]   Gascon J, Bern C, Pinazo MJ. Chagas disease in Spain, the United States and other non-endemic countries. Acta Trop 2010; 115(1-2): 22-7.
        [http://dx.doi.org/10.1016/j.actatropica.2009.07.019] [PMID: 19646412]

[83]   Ribeiro AL, Nunes MP, Teixeira MM, Rocha MO. Diagnosis and management of Chagas disease and cardiomyopathy. Nat Rev Cardiol 2012; 9(10): 576-89.
        [http://dx.doi.org/10.1038/nrcardio.2012.109] [PMID: 22847166]

[84]   Goldenberg RC, Jelicks LA, Fortes FS, *et al.* Bone marrow cell therapy ameliorates and reverses chagasic cardiomyopathy in a mouse model. J Infect Dis 2008; 197(4): 544-7.
        [http://dx.doi.org/10.1086/526793] [PMID: 18237267]

[85]   Jones EM, Colley DG, Tostes S, Lopes ER, Vnencak-Jones CL, McCurley TL. Amplification of a Trypanosoma cruzi DNA sequence from inflammatory lesions in human chagasic cardiomyopathy. Am J Trop Med Hyg 1993; 48(3): 348-57.
        [PMID: 8470772]

[86]   Bellotti G, Bocchi EA, de Moraes AV, *et al.* *In vivo* detection of Trypanosoma cruzi antigens in hearts of patients with chronic Chagas ?(tm) heart disease. Am Heart J 1996; 131(2): 301-7.
        [http://dx.doi.org/10.1016/S0002-8703(96)90358-0] [PMID: 8579025]

[87]   Zhang L, Tarleton RL. Parasite persistence correlates with disease severity and localization in chronic Chagas ?(tm) disease. J Infect Dis 1999; 180(2): 480-6.
        [http://dx.doi.org/10.1086/314889] [PMID: 10395865]

[88]   Basquiera AL, Sembaj A, Aguerri AM, *et al.* Risk progression to chronic Chagas cardiomyopathy: influence of male sex and of parasitaemia detected by polymerase chain reaction. Heart 2003; 89(10): 1186-90.
        [http://dx.doi.org/10.1136/heart.89.10.1186] [PMID: 12975414]

[89]   Koberle F. Chagas' disease and Chagas ?(tm) syndromes: the pathology of American trypanosomiasis. Adv Parasitol 1968; 6: 63-116.
        [http://dx.doi.org/10.1016/S0065-308X(08)60472-8] [PMID: 4239747]

[90]   Soares MB, Lima RS, Rocha LL, *et al.* Transplanted bone marrow cells repair heart tissue and reduce myocarditis in chronic chagasic mice. Am J Pathol 2004; 164(2): 441-7.
        [http://dx.doi.org/10.1016/S0002-9440(10)63134-3] [PMID: 14742250]

[91]   Goldenberg RC, Jelicks LA, Fortes FS, *et al.* Bone marrow cell therapy ameliorates and reverses chagasic cardiomyopathy in a mouse model. J Infect Dis 2008; 197(4): 544-7.
        [http://dx.doi.org/10.1086/526793] [PMID: 18237267]

[92]   Bartholomew A, Sturgeon C, Siatskas M, *et al.* Mesenchymal stem cells suppress lymphocyte proliferation *in vitro* and prolong skin graft survival *in vivo.* Exp Hematol 2002; 30(1): 42-8.
        [http://dx.doi.org/10.1016/S0301-472X(01)00769-X] [PMID: 11823036]

[93]   Krampera M, Glennie S, Dyson J, *et al.* Bone marrow mesenchymal stem cells inhibit the response of naive and memory antigen-specific T cells to their cognate peptide. Blood 2003; 101(9): 3722-9.
        [http://dx.doi.org/10.1182/blood-2002-07-2104] [PMID: 12506037]

[94]   Reis MM, Higuchi MdeL, Benvenuti LA, *et al.* An in situ quantitative immunohistochemical study of

cytokines and IL-2R+ in chronic human chagasic myocarditis: correlation with the presence of myocardial Trypanosoma cruzi antigens. Clin Immunol Immunopathol 1997; 83(2): 165-72.
[http://dx.doi.org/10.1006/clin.1997.4335] [PMID: 9143377]

[95]   Jiang Y, Jahagirdar BN, Reinhardt RL, *et al.* Pluripotency of mesenchymal stem cells derived from adult marrow. Nature 2002; 418(6893): 41-9.
[http://dx.doi.org/10.1038/nature00870] [PMID: 12077603]

[96]   Macambira SG, Vasconcelos JF, Costa CR, *et al.* Granulocyte colony-stimulating factor treatment in chronic Chagas disease: preservation and improvement of cardiac structure and function. FASEB J 2009; 23(11): 3843-50.
[http://dx.doi.org/10.1096/fj.09-137869] [PMID: 19608624]

[97]   Vasconcelos JF, Souza BS, Lins TF, *et al.* Administration of granulocyte colony-stimulating factor induces immunomodulation, recruitment of T regulatory cells, reduction of myocarditis and decrease of parasite load in a mouse model of chronic Chagas disease cardiomyopathy. FASEB J 2013; 27(12): 4691-702.
[http://dx.doi.org/10.1096/fj.13-229351] [PMID: 23964077]

[98]   Larocca TF, Souza BS, Silva CA, *et al.* Transplantation of adipose tissue mesenchymal stem cells in experimental chronic chagasic cardiopathy. Arq Bras Cardiol 2013; 100(5): 460-8.
[PMID: 23568098]

[99]   Silva DN, de Freitas Souza BS, Azevedo CM, *et al.* Intramyocardial transplantation of cardiac mesenchymal stem cells reduces myocarditis in a model of chronic Chagas disease cardiomyopathy. Stem Cell Res Ther 2014; 5(4): 81.
[http://dx.doi.org/10.1186/scrt470] [PMID: 24984860]

[100] Machado FS, Jelicks LA, Kirchhoff LV, *et al.* Chagas heart disease: report on recent developments. Cardiol Rev 2012; 20(2): 53-65.
[PMID: 22293860]

[101] Vilas-Boas F, Feitosa GS, Soares MB, *et al.* Bone marrow cell transplantation to the myocardium of a patient with heart failure due to Chagas ?(tm) disease. Arq Bras Cardiol 2004; 82(2): 185-187, 181-184.
[PMID: 15042255]

[102] Vilas-Boas F, Feitosa GS, Soares MB, *et al.* [Early results of bone marrow cell transplantation to the myocardium of patients with heart failure due to Chagas disease]. Arq Bras Cardiol 2006; 87(2): 159-66.
[PMID: 16951834]

[103] Soares MB, Santos RR. Current status and perspectives of cell therapy in Chagas disease. Mem Inst Oswaldo Cruz 2009; 104 (Suppl. 1): 325-32.
[http://dx.doi.org/10.1590/S0074-02762009000900043] [PMID: 19753492]

[104] Ribeiro Dos Santos R, Rassi S, Feitosa G, *et al.* Cell therapy in Chagas cardiomyopathy (Chagas arm of the multicenter randomized trial of cell therapy in cardiopathies study): a multicenter randomized trial. Circulation 2012; 125(20): 2454-61.
[http://dx.doi.org/10.1161/CIRCULATIONAHA.111.067785] [PMID: 22523306]

[105] Zhang J, Wilson GF, Soerens AG, *et al.* Functional cardiomyocytes derived from human induced

pluripotent stem cells. Circ Res 2009; 104(4): e30-41.
[http://dx.doi.org/10.1161/CIRCRESAHA.108.192237] [PMID: 19213953]

[106] Soares MB, Santos RR. Current status and perspectives of cell therapy in Chagas disease. Mem Inst Oswaldo Cruz 2009; 104 (Suppl. 1): 325-32.
[http://dx.doi.org/10.1590/S0074-02762009000900043] [PMID: 19753492]

[107] Cheng AS, Yau TM. Paracrine effects of cell transplantation: strategies to augment the efficacy of cell therapies. Semin Thorac Cardiovasc Surg 2008; 20(2): 94-101.
[http://dx.doi.org/10.1053/j.semtcvs.2008.04.003] [PMID: 18707640]

[108] Robey TE, Saiget MK, Reinecke H, Murry CE. Systems approaches to preventing transplanted cell death in cardiac repair. J Mol Cell Cardiol 2008; 45(4): 567-81.
[http://dx.doi.org/10.1016/j.yjmcc.2008.03.009] [PMID: 18466917]

[109] Rink JC. Stem cell systems and regeneration in planaria. Dev Genes Evol 2013; 223(1-2): 67-84.
[http://dx.doi.org/10.1007/s00427-012-0426-4] [PMID: 23138344]

[110] Gentile L, CebriA F, Bartscherer K. The planarian flatworm: an *in vivo* model for stem cell biology and nervous system regeneration. Dis Model Mech 2011; 4(1): 12-9.
[http://dx.doi.org/10.1242/dmm.006692] [PMID: 21135057]

[111] Collins JJ III, Wang B, Lambrus BG, Tharp ME, Iyer H, Newmark PA. Adult somatic stem cells in the human parasite Schistosoma mansoni. Nature 2013; 494(7438): 476-9.
[http://dx.doi.org/10.1038/nature11924] [PMID: 23426263]

[112] Wang B, Collins JJ III, Newmark PA. Functional genomic characterization of neoblast-like stem cells in larval Schistosoma mansoni. eLife 2013; 2: e00768.
[http://dx.doi.org/10.7554/eLife.00768] [PMID: 23908765]

[113] Fenwick A. The global burden of neglected tropical diseases. Public Health 2012; 126(3): 233-6.
[http://dx.doi.org/10.1016/j.puhe.2011.11.015] [PMID: 22325616]

[114] van der Werf MJ, de Vlas SJ, Brooker S, *et al.* Quantification of clinical morbidity associated with schistosome infection in sub-Saharan Africa. Acta Trop 2003; 86(2-3): 125-39.
[http://dx.doi.org/10.1016/S0001-706X(03)00029-9] [PMID: 12745133]

[115] Walker AJ. Insights into the functional biology of schistosomes. Parasit Vectors 2011; 4: 203.
[http://dx.doi.org/10.1186/1756-3305-4-203] [PMID: 22013990]

[116] Basch PF. Schistosomes: Development, Reproduction, and Host Relations. Oxford University Press 1991.

[117] Newmark PA, *et al.* Stash of Stem Cells Found in a Human Parasite 2013.

[118] Collins JJ III, Newmark PA. It ?(tm)s no fluke: the planarian as a model for understanding schistosomes. PLoS Pathog 2013; 9(7): e1003396.
[http://dx.doi.org/10.1371/journal.ppat.1003396] [PMID: 23874195]

[119] Wagner DE, Ho JJ, Reddien PW. Genetic regulators of a pluripotent adult stem cell system in planarians identified by RNAi and clonal analysis. Cell Stem Cell 2012; 10(3): 299-311.
[http://dx.doi.org/10.1016/j.stem.2012.01.016] [PMID: 22385657]

[120] Onal P, GrA1/4n D, Adamidi C, *et al.* Gene expression of pluripotency determinants is conserved

between mammalian and planarian stem cells. EMBO J 2012; 31(12): 2755-69.
[http://dx.doi.org/10.1038/emboj.2012.110] [PMID: 22543868]

[121]   Das B, Kashino SS, Pulu I, *et al.* CD271(+) bone marrow mesenchymal stem cells may provide a niche for dormant Mycobacterium tuberculosis. Sci Transl Med 2013; 5(170): 170ra13.
[http://dx.doi.org/10.1126/scitranslmed.3004912] [PMID: 23363977]

[122]   Eckert J, Deplazes P. Biological, epidemiological, and clinical aspects of echinococcosis, a zoonosis of increasing concern. Clin Microbiol Rev 2004; 17(1): 107-35.
[http://dx.doi.org/10.1128/CMR.17.1.107-135.2004] [PMID: 14726458]

[123]   Brehm K. The role of evolutionarily conserved signaling systems in *Echinococcus multilocularis* development and host-parasite interaction. Med Microbiol Immunol (Berl) ? ? ? ; 201(199): 247-59.

[124]   Brehm K. *Echinococcus multilocularis* as an experimental model in stem cell research and molecular host-parasite interaction. Parasitology 2010; 137(3): 537-55.
[http://dx.doi.org/10.1017/S0031182009991727] [PMID: 19961652]

[125]   Spiliotis M, Lechner S, Tappe D, Scheller C, Krohne G, Brehm K. Transient transfection of *Echinococcus multilocularis* primary cells and complete *in vitro* regeneration of metacestode vesicles. Int J Parasitol 2008; 38(8-9): 1025-39.
[http://dx.doi.org/10.1016/j.ijpara.2007.11.002] [PMID: 18086473]

[126]   Koziol U, Rauschendorfer T, Zanon RodrA-guez L, Krohne G, Brehm K. The unique stem cell system of the immortal larva of the human parasite *Echinococcus multilocularis.* Evodevo 2014; 5(1): 10.
[http://dx.doi.org/10.1186/2041-9139-5-10] [PMID: 24602211]

[127]   Schubert A, Koziol U, Cailliau K, Vanderstraete M, Dissous C, Brehm K. Targeting *Echinococcus multilocularis* stem cells by inhibition of the Polo-like kinase EmPlk1. PLoS Negl Trop Dis 2014; 8(6): e2870.
[http://dx.doi.org/10.1371/journal.pntd.0002870] [PMID: 24901228]

[128]   Valent P. Cytokines involved in growth and differentiation of human basophils and mast cells. Exp Dermatol 1995; 4(4 Pt 2): 255-9.
[http://dx.doi.org/10.1111/j.1600-0625.1995.tb00254.x] [PMID: 8528598]

[129]   Metcalf D. Hematopoietic cytokines. Blood 2008; 111(2): 485-91.
[http://dx.doi.org/10.1182/blood-2007-03-079681] [PMID: 18182579]

[130]   Wehrle-Haller B. The role of Kit-ligand in melanocyte development and epidermal homeostasis. Pigment Cell Res 2003; 16(3): 287-96.
[http://dx.doi.org/10.1034/j.1600-0749.2003.00055.x] [PMID: 12753403]

[131]   Copley MR, Eaves CJ. Developmental changes in hematopoietic stem cell properties. Exp Mol Med 2013; 45: e55.
[http://dx.doi.org/10.1038/emm.2013.98] [PMID: 24232254]

[132]   Rönnstrand L. Signal transduction *via* the stem cell factor receptor/c-Kit. Cell Mol Life Sci 2004; 61(19-20): 2535-48.
[http://dx.doi.org/10.1007/s00018-004-4189-6] [PMID: 15526160]

[133]   Seita J, Weissman IL. Hematopoietic stem cell: self-renewal *versus* differentiation. Wiley Interdiscip Rev Syst Biol Med 2010; 2(6): 640-53.

[http://dx.doi.org/10.1002/wsbm.86] [PMID: 20890962]

[134] Zhang CC, Lodish HF. Cytokines regulating hematopoietic stem cell function. Curr Opin Hematol 2008; 15(4): 307-11.
[http://dx.doi.org/10.1097/MOH.0b013e3283007db5] [PMID: 18536567]

[135] Neill DR, McKenzie AN. Nuocytes and beyond: new insights into helminth expulsion. Trends Parasitol 2011; 27(5): 214-21.
[http://dx.doi.org/10.1016/j.pt.2011.01.001] [PMID: 21292555]

[136] Newlands GF, Coulson PS, Wilson RA. Stem cell factor dependent hyperplasia of mucosal-type mast cells but not eosinophils in Schistosoma mansoni-infected rats. Parasite Immunol 1995; 17(11): 595-8.
[http://dx.doi.org/10.1111/j.1365-3024.1995.tb01003.x] [PMID: 8817606]

[137] Urb M, Sheppard DC. The role of mast cells in the defence against pathogens. PLoS Pathog 2012; 8(4): e1002619.
[http://dx.doi.org/10.1371/journal.ppat.1002619] [PMID: 22577358]

[138] Furuta T, Kikuchi T, Iwakura Y, Watanabe N. Protective roles of mast cells and mast cell-derived TNF in murine malaria. J Immunol 2006; 177(5): 3294-302.
[http://dx.doi.org/10.4049/jimmunol.177.5.3294] [PMID: 16920970]

[139] Assawachananont M, Mandai J, Okamoto S, *et al.* Transplantation of Embryonic and Induced Pluripotent Stem Cell-Derived 3D Retinal Sheets into Retinal Degenerative. Mice Stem Cell Reports 2014; 2(5): 662-74.
[http://dx.doi.org/10.1016/j.stemcr.2014.03.011] [PMID: 24936453]

[140] Lantz CS, Boesiger J, Song CH, *et al.* Role for interleukin-3 in mast-cell and basophil development and in immunity to parasites. Nature 1998; 392(6671): 90-3.
[http://dx.doi.org/10.1038/32190] [PMID: 9510253]

[141] Donaldson LE, Schmitt E, Huntley JF, Newlands GF, Grencis RK. A critical role for stem cell factor and c-kit in host protective immunity to an intestinal helminth. Int Immunol 1996; 8(4): 559-67.
[http://dx.doi.org/10.1093/intimm/8.4.559] [PMID: 8671643]

[142] Yuan Q, Gurish MF, Friend DS, Austen KF, Boyce JA. Generation of a novel stem cell factor-dependent mast cell progenitor. J Immunol 1998; 161(10): 5143-6.
[PMID: 9820483]

[143] Oliveira SH, Taub DD, Nagel J, *et al.* Stem cell factor induces eosinophil activation and degranulation: mediator release and gene array analysis. Blood 2002; 100(13): 4291-7.
[http://dx.doi.org/10.1182/blood.V100.13.4291] [PMID: 12453875]

[144] Newlands GF, Miller HR, MacKellar A, Galli SJ. Stem cell factor contributes to intestinal mucosal mast cell hyperplasia in rats infected with Nippostrongylus brasiliensis or Trichinella spiralis, but anti-stem cell factor treatment decreases parasite egg production during N brasiliensis infection. Blood 1995; 86(5): 1968-76.
[PMID: 7544650]

[145] Puddington L, Olson S. Interactions between stem cell factor and c-Kit are required for intestinal immune system homeostasis. Immunity 1994; 1(9): 733-9.
[http://dx.doi.org/10.1016/S1074-7613(94)80015-4] [PMID: 7534619]

[146] Yazdanbakhsh M. IgE, eosinophils and mast cells in helminth infections. Ned Tijdschr Klin Chem 1996; 21: 213-6.

[147] Hepworth MR, Maurer M, Hartmann S. Regulation of type 2 immunity to helminths by mast cells. Gut Microbes 2012; 3(5): 476-81.
[http://dx.doi.org/10.4161/gmic.21507] [PMID: 22892692]

SUBJECT INDEX

www.ingramcontent.com/pod-product-compliance
Lightning Source LLC
Chambersburg PA
CBHW050822220326
41598CB00006B/288